science and the modern world

contributors

Ansley J. Coale
Theodosius Dobzhansky
Bentley Glass
Carl C. Kiess
James McCormack
Ernan McMullin
G. C. McVittie
Philip M. Morse
E. R. Piore
Arthur E. Ruark

science and the modern world

One of a series of lectures presented at Georgetown University, Washington, D. C. on the occasion of its 175th Anniversary, October 1963 to May 1964

Edited by
JACINTO STEINHARDT

PLENUM PRESS · NEW YORK · 1966

Library of Congress Catalog Card Number 66-19930

Plenum Press
A Division of Plenum Publishing Corporation
227 West 17 Street, New York, N. Y. 10011

Printed in the United States of America

Introduction

During the period September 1963 to December 1964, Georgetown University commemorated the 175th anniversary of its founding by presenting a series of special lectures in the various disciplines appropriate to a university. Among the subjects covered, in separate and coherent series, were the humanities, the arts, medicine, law, and religion. One of the series, entitled as a group, "Science and Society," is represented by this book.

The ten papers presented here must be looked at in context: an effort was made to present to a largely lay audience a cross section of those aspects of recent advances in science and technology that were considered to have the most striking or inclusive effects on man's view of his world, on desirable public policy, and on man's anticipations as to how further advances and applications of these advances were likely to change his environment.

Clearly, so ambitious an undertaking could not be carried out within the compass of ten brief lectures. Within a general structure, it was necessary to use a sampling technique. The general plan was adopted of having three lectures representative of the philosophical sector, three of the public policy structure, and three which would partake of the nature of extrapolations into the future. A tenth lecture was later added which could be considered to fit equally well into the second or third of these categories.

Within the philosophical sector, it was considered that the great effect of science during the immediate past had been in the fields of astronomy or astrophysics, and biology. It was also considered appropriate to balance accounts of these fields with a prudent appreciation of the limitations of science. At a time when even man's ethical and aesthetic value systems have been attributed to the effects of natural selection, it seemed important to be reminded of the extent to which value systems and the affective contents of experience cannot rest on a material base, and as well, of the limits that such barriers as considerations of attainable energies set beyond which scientific experiment cannot possibly go.

In the public policy sector, lectures on the practical uses of atomic energy, and on population trends and population control, were joined to a general discussion of the relationship of government sponsorship to scientific exploration.

In the realm of extrapolations to the future the talks were planned about materials and instrumentation, the development of the computer sciences, and biology and medicine, as the centers from which the most drastic general effects would come.

It is one thing to conceive a grand outline. It is much more difficult to bring it to fruition. The list of the titles of the lectures, and of the distinguished speakers who spoke on each of the topics, gives a clear indication of what was intended. A perusal of the texts will show, however, that many of the speakers either restricted themselves to particular aspects of the topics suggested by the titles or, in one or two cases, departed from them substantially. While the grand design may have suffered from these changes, they added depth where they reduced scope, or introduced an unexpected emphasis on timeliness which was very appropriate in the year of the anniversary. Thus, to cite a single example, Professor Dobzhansky's lecture dwells at some length on aspects of racism, which was very much on the minds of all audiences at the time of his lecture.

The thanks of the University are due to each of the ten speakers who made this series the success that it was. All of them are active men with heavy demands on their time. The delay in publishing this volume is a natural result of that fact. The material is of such a nature, however, that its timeliness has not been impaired by the extra year that has elapsed from the date when the earliest publication could have been expected. Thanks are due also to Mrs. Harriet Hubbard for her patient care in preparing the manuscripts for publication.

J.S.

ADDENDUM

"The Revolution in Biology and Medicine," by Bentley Glass, pp. 187 to 205, is based in part on two earlier articles by Dr. Glass, portions of which are here reprinted with the permission of the copyright owners, which is gratefully acknowledged:

Contents

New Developments in Our Knowledge of the Universe

G. C. McVittie

G. C. McVittie *was educated at the University of Edinburgh, (M.A. 1927) and the University of Cambridge (Ph.D. 1930). After holding posts in the Universities of Leeds, Edinburgh, and Liverpool, he was appointed Reader in Mathematics at King's College, (University of London) in 1936 and Professor of Mathematics at Queen Mary College (University of London) in 1948. During the war years (1939-1945) McVittie was engaged in meteorological work and was made an Officer of the Order of British Empire in 1946 in recognition of his services. Since 1952, he has been Professor of Astronomy and Head of Department at the University of Illinois in Urbana. He has written extensively on the applications of Einstein's general relativity to astronomical problems, in particular, on the cosmological problem. Two recent books of his are "General Relativity and Cosmology" and "Fact and Theory in Cosmology." He has been Secretary of the American Astronomical Society since 1961.*

New Developments in Our Knowledge of the Universe

RECENT DEVELOPMENTS in cosmology are the result of the impact of a mixture of disciplines on astronomy. It is the introduction of radio astronomy that is responsible for the most remarkable of these developments. And the prosecution of radio astronomy entails electrical engineering, the use of bulldozers, the methods of dealing with mud and electronic physics as well as astronomy.

I shall assume that you are familiar with the term "galaxy" and that perhaps you know what the spectrum of a luminous object is. But in case you have forgotten, I will remind you that if white light is passed through a prism, or some equivalent device, a band of colored light is obtained, with violet or ultraviolet at one end, blue, green, and yellow in between, and red at the other end. When the light comes from an astronomical object, such as an entire galaxy or an individual star, certain black lines are also present and they are known as the spectral lines, which are, so to speak, the footprints left by the chemical elements found in the outer layers of the luminous object. In some spectra however there are also emission lines—that is to say brilliant lines—which arise because a certain type of atom is very common in the outer layers of the object. These atoms are themselves emitting light and, therefore, they produce an enhancement of the spectrum at that particular point.

A very important datum for cosmology is the so-called red-shift, namely, a displacement of all the spectral lines towards the red end of the spectrum. This occurs in a way which compels us to say that the displacement means that the galaxies are moving away from our own. It is a Doppler shift. The method of measuring the red-shift is the following: The wavelength, in the laboratory, of a spectral line being λ, the Doppler shift increases the wavelength to $\lambda + \delta\lambda$. Then the red-shift, which is usually noted in the literature by the letter z, is the ratio of the increment of wavelength $\delta\lambda$ to the original or laboratory wavelength λ, so that $z = \delta\lambda/\lambda$.

This is what astronomers observe, though they like to describe it, just to confuse the issue, by multiplying z by the velocity of light c. Thus there is obtained what has been called a ficitious or representative velocity. I shall avoid this device and use z instead. For the nearer clusters of galaxies, the Virgo cluster for instance, z is small, equal to about 0.004. The largest red-shift which has been measured, in an object to which I shall presently refer, is as large as 0.46. There is no theoretical reason why red-shifts equal to 1, or greater than 1, should not occur, and presumably some day objects with red-shifts greater than 0.46 will be identified and they will no doubt be galaxies of some kind.

In a general way it may be said that the larger the red-shift, the farther away is the object in question though distance is not simply proportional to red-shift. Up to May 1960, the largest red-shift that had been measured had been observed by spectroscopic means in the Hydra cluster of galaxies, and the amount was a little larger than 0.2. But in May 1960—you are no doubt familiar with this charming story—Dr. Minkowski at Mount Wilson and Palomar, who was about to retire, had his last 10-day run on the 200-in. telescope, and he decided to devote it to the study of a certain object he believed would be interesting from the red-shift point of view. This object was suggested to him by the work of radio astronomers. It is No. 295 in the third Cambridge catalog, and is therefore known as 3C295. Minkowski, talking to the radio astronomers at the California Institute of Technology, concluded that 3C295 was an intrinsically strong radio source, even if it did not appear strong in terrestrial radio telescopes. The reason why it did not appear strong, Minkowski conjectured, was that it was very remote.

The radio astronomers had fixed the position of 3C295 to within a small area of the sky. The area was not so large that the 200-in. optical telescope, with its small field of view, could not examine it in a reasonable time. The first thing Minkowski noticed was that 3C295 appeared to be the brightest galaxy of a group of about 60 galaxies. A spectrum with a 9-hr exposure showed one single bright emission line, and very little else. Minkowski decided that this emission line was the well-known line of singly-ionized oxygen. But in order to achieve this interpretation it was necessary to suppose that the red-shift was 0.46. Admittedly, even by cosmological standards, this result was a trifle uncertain, to say the least. There was one single identified spectral line and, on this basis, a red-shift of more than double what was known before had been obtained. However, there were the other members of the cluster to which 3C295 appeared to belong, and therefore Minkowski asked his colleague Dr. W. A. Baum to study some of the other members of the cluster by a different technique which I will not describe. And, happily, as a result of Baum's studies on, I believe, three more of the galaxies of the cluster it was revealed that indeed the whole cluster had a red-shift of 0.46. By the aid of radio astronomy a great leap forward into the remoteness of space had thus been taken.

There is a much closer object known as Cygnus A which is also a radio source and is a galaxy of the same general type as 3C295. And from the optical point of view there is no difficulty in saying that both objects are galaxies, though perhaps one might think that Cygnus A is not a very neat and orderly kind of galaxy, but one in which unusual events are taking place.

Radio astronomy has done more than this, however, during the past year by drawing attention to very much more remarkable kinds of objects. These objects, of which four or five are certainly known and many more are no doubt waiting to be discovered, are strong radio sources. But they possess the curious feature that, unlike 3C295, Cygnus A or many of the other known moderate or strong radio sources, they are found in isolation and are not members of clusters of galaxies. Moreover, their optical images on the photographic plate look just like the images of stars.

One of these objects is 3C273 which, in a photograph, looks like a star, except that it does have a narrow filament of gaseous material apparently stretching away from the stellar-like main body. The optical spectrum of the latter has been interpreted by Maarten Schmidt, an interpretation which leads to a red-shift of 0.158. This would correspond to a distance of the order of 1.6 billion light years and this in turn indicates that the central body is 40 to 100 times more luminous than 3C295 or Cygnus A, themselves no mean radiators of light. Moreover, Harlan Smith has found that 3C273 had been photographed on plates at Harvard for 70 years or more. Examination of this long series of plates showed that the emission of light was variable with roughly a 10-yr period. Now and again, notably in 1927-29, the object suddenly increased in luminosity. At the maximum of such a flash, 3C273 is 1600 times as bright as the brightest supernova. The flashes last from 2.8 hours to 12 days. Another remarkable fact about 3C273 is that its radio emission comes from two separate centers; one is close to the main body, the other lies near the far end of the jet of gas. This was established by Hazard and his colleagues in an very elegant lunar occultation observation carried out in Australia.

Another object of the class is 3C48, whose optical spectrum is interpretable if a red-shift of 0.37 is assumed. The distance is of the order of 3.6 billion light years and again we have a very luminous object which is also a strong radio source. There are clear indications that its emission of light is variable. Near the stellar-like main body there is some nebulosity which, by itself, is more luminous than are many entire galaxies which are found in clusters. Other objects of the class are 3C196 and 3C286, though less is known about them at the moment.

The intriguing question is now: What are these objects and what is going on within them? How can bodies of such apparently small size emit so much energy in the form of light and also of radio waves? And, in addition, what could cause the variability in the emission of light? I do not know what the answers to these questions are, at present, and though some of my more quick-witted colleagues are suggesting answers, I think we all ought to do a good deal more work before we commit ourselves!

I have heard some of my own contemporaries saying that these objects represent one of the biggest discoveries that has ever been made in astronomy. The implication is that everybody should now rush and study the problem. However, astronomers—and indeed scientists in general—are rather like women when it comes to fashion: everybody must be doing the fashionable thing at the same time. But in spite of the fact that I personally am extremely interested in these objects and would very much like to know what they are, I do not think that we should all rush like a pack of hounds after this one particular kind of investigation! The analogy, of course, in stellar theory would be to say that every astronomer should abandon studies of all stars except supernovae and novae, two kinds of stars that are blowing up and generally showing great signs of commotion. I am sure much would be discovered if more was known about supernovae. But this does not mean, I think, that everyone should rush exclusively into the study of these particular stars.

It seems to me that these objects draw attention to something which has not hitherto been considered. Astronomers are familiar with the fact that an individual star can blow up, which leads to the problem of the nature of a supernova. But I think that astronomers had imagined that such disturbances were restricted to individual stars; whereas now it seems fairly clear that this is not the case. An entire galaxy, or at any rate a large fraction of one, can behave in this explosive manner. In a sense, this is not new. Radio astronomers have drawn attention to certain other objects in which explosive events have occurred. For example, the member of the Virgo cluster of galaxies known as NGC 4486, seems to contain a jet of hydrogen gas which is somehow being ejected from the center of this galaxy. This phenomenon has been known for a number of years and though it was regarded as very remarkable it was perhaps unique. However, more recently, Sandage and Lynds have studied the very irregular galaxy M 82. They show that from the central region there is a great complex of filaments of hydrogen gas going outward more or less at right angles to the central plane of the galaxy. They have been able to measure the rate at which the jets of hydrogen are being ejected, and they reckon that if the speeds now observed have persisted for some time past, only 1.5 million years are required to bring all the gas back to the middle of the galaxy. Presumably what has happened in the past is an ejection of hydrogen gas from the center of the galaxy, of total mass about 5 million solar masses.

I suppose one might hazard the guess that these two objects reveal a very feeble version of what is going on in 3C273 and 3C48. It has been suggested that, in 3C48 and 3C273, we are watching the aftermath of the collapse to a very small radius of a great mass of gas. The collapse produced an enormous amount of gravitational potential energy which is now being liberated, by being converted into radio or optical waves. This picture leads us to contemplate the possibility that great masses—by which I mean of the order of some millions of times the solar mass—can operate in an organized condition of implosion followed by explosion. How a state of instability which causes an amount of material, enormous even by astronomical standards, to shrink to a radius of almost nothing is produced, and how this is followed by an explosion, are wide-open problems at present.

So much for these problems which the interplay of radio and optical astronomy has brought to light in the last few years. I should like now to say something about another trouble which the radio astronomers have brought into this world. This is the question of the counts of extragalactic radio sources.

Let me first tell you something about the counts of optical galaxies. Some region of the sky, large enough and clear of obscuring matter, is selected that will, it is hoped, provide a good sample of the whole celestial sphere. Then all the galaxies that are seen on photographic plates of the same exposure time and taken with the same instrument, are counted. In principle, all galaxies not fainter than apparent magnitude, say, 13, are counted; then the count is repeated to apparent magnitude 15, then to 17, and so on down to 21.

Data of this type, in which the number of galaxies in a given region of the sky is counted to a series of limiting magnitudes, are called number counts. A number count thus reveals how the number of galaxies increases as their brightness gets fainter and fainter, in other words, as galaxies that are further and further away are included in the count.

This kind of work was done by Hubble in the 1930's. It has been repeated more recently only for the brighter limiting magnitudes with better observing techniques. But as far as the depth of the universe goes, the only data are those of Hubble. Astronomers like to criticize Hubble's work, saying that his limiting magnitudes are not good and that his counts contain this or that defect. A theoretician like myself tends to reply that if Hubble's work is so bad and cannot be relied on, why is the work not repeated, and the true facts revealed? The answer I am given is that a repetition would take a long time. Therefore we must either accept the number counts or we confess our complete ignorance and sit back until somebody does have the patience to repeat Hubble's work. If we accept the counts we are agreeing that Hubble was one of the most expert observers of our time, and that he could not have made errors so wild, so terrible, that his results are nonsensical.

Suppose then that we do accept Hubble's data and compare what is observed with the predictions of the simplest theory that can be manufactured. Such a theory is based on the following three ideas: First, assume that all galaxies are intrinsically equally powerful emitters of light. Second, assume that all galaxies are at mutual rest not only with respect to our galaxy but with respect to each other; in other words, ignore the red-shift phenomenon for a moment. And third, assume that the Euclidean space of classical physics ought to be used and that the galaxies are scattered evenly throughout this classical space.

The result of these three assumptions is the magic "minus three-halves law," which gives a relation between the expected number of galaxies and each limit of brightness. The observed counts do fit this law fairly well if the counts only include bright galaxies to about apparent magnitude 12 or 13. In fact, they fit suprisingly well. But when fainter galaxies are included, the "minus three-halves law" begins to give higher numbers of galaxies than are observed. And the fainter you go, the worse the discrepancy becomes.

It is very easy to correct the trouble by looking back at the initial assumptions made in the simple theory, especially at the most glaringly inaccurate one, namely that all galaxies are at relative rest. In other words, the simple theory is modified by allowing for the expansion phenomenon to which we attribute the red-shift. Using the expanding universe theory, and still keeping to uniformity as it is defined in that theory, it is easy to show why the "minus three-halves law" gives too many galaxies. If the expansion is properly allowed for, agreement is obtained between the observations of optical counts and the theoretical predictions.

This is very nice, and the pleasant situation continued

until radio astronomy came along. In Cambridge, England, and in Australia, two different instruments with different methods of observing, were employed to count extragalactic radio sources. I need not go into the technicalities; suffice it to say that a radio apparent magnitude can be defined entirely comparable to the optical apparent magnitude, and then the counting can be done in much the same way as in the optical case. But when these observations are compared with the "minus three-halves law" it is found to be inadequate. Worse still, inclusion of the expansion effect does not remedy the trouble. It is found that there are far too many radio galaxies as the counts proceed into the remoteness of space.

Various ways have been proposed for getting out of this difficulty. The first one is a favored one in cosmology: if the observations do not fit a theory there must be something wrong with the data; discard them. In this case there is perhaps more reason for doing so than usual because there are many difficulties about the data; they are not even as accurate as Hubble's presumably were. There are many features of the observations which would make a radio astronomer suspect the presence of systematic error. This might account for the strange result. However, suppose that the data are accepted and that we take another look at the theory. What about the hypothesis that every radio source is intrinsically equally powerful? Suppose that we put into the theory the luminosity function, which is a device that allows for variations in intrinsic radio power. A student of mine, R. C. Roeder, recently solved this problem. It is a complicated matter if it is done without cheating. By this I mean that the expansion effect must be included from the start, not added later on, like an excrescence, to a statical treatment. The luminosity function produces some improvement but not enough. The discrepancy is still there. There are still too many radio sources.

There are more extreme suggestions. Hoyle and Narlikar, for instance, have a theory in which, in effect, they have abandoned the hypothesis of uniformity. At least they abandon it in one part of the theory, it seems to me, and retain it in another part. It is a very interesting line of approach, because I think if one were to begin again from the beginning and do everything in a nonuniform universe, which would be a terrible job mathematically, there might be some hope of solving the problem. I am personally inclined to interpret these matters a little more at their face value on the grounds that radio astronomy has revealed so much that is odd that another odd feature may be thrown in for good meansre. I would say that the observations tell us what is in fact the case; namely, that there were intrinsically more radio sources per unit volume in the past than there are now in our neighborhood. Remoteness in space means also remoteness in time because radio waves travel at a finite speed. Our neighborhood reveals the state of affairs prevalent at a moment some hundreds of millions of years, or even thousands of millions of years, later in the history of the universe, than the one observed in the counts of remote sources. Therefore, there are too few surviving radio galaxies in our neighborhood compared with the numbers found when we look back into the remote past. The argument implies, of course, that strong radio emission by a galaxy is a short-lived phenomenon.

There is another variant of this theory which also does the trick. Instead of saying that there were more radio sources per unit volume in the past, one keeps the number constant but assumes that they were more powerful radiators in the past. These are simple-minded interpretations. I have nothing to offer as to why there should have been more radio galaxies per unit volume on the average, or why they should have been brighter in the remote past than there are now. One would have to know what makes a galaxy a radio source to begin with. And is there any evidence that strong radio emission is a short-lived phenomenon when it occurs?

Once again, in the counts problem, we are confronted by a puzzle. The solution may, of course, be simply a matter of getting better observations. Perhaps if we are lucky at

Illinois, where we have constructed a radio telescope specially adapted to the counts problem, we may be able to throw some light on it. Only time can tell.

Every time a new method of observing is invented, an immense number of interesting problems is thereby produced. Radio astronomy has succeeded in raising as many problems in cosmology—in fact more problems in cosmology—than it has so far solved. And, for my part I say, long may this continue!

Biological Evolution and Human Equality

Theodosius Dobzhansky

***Theodosius Dobzhansky**, of the Rockefeller University, was born in Russia in 1900. He was graduated from the University of Kiev in 1921. After a teaching and research career in Russia he came to the United States in 1928 as a Fellow of the International Education Board of the Rockefeller Foundation to work with T. H. Morgan, the principal creator of modern genetics, at Columbia University. He followed Morgan to the California Institute of Technology a year or so later, and stayed there until 1940, when he returned to Columbia University. In 1962 he joined the Rockefeller University. He is the author of numerous articles and books, including "Mankind Evolving," published in 1963. His book, "Genetics and the Origin of Species," was a landmark in the development of both evolutionary theory and genetics. He is a member of learned societies, and the winner of various prizes and has been awarded a number of honorary degrees. His most recent award was the National Medal of Science, in 1964.*

Biological Evolution and Human Equality

THE CRUCIAL FACT of our age is that people almost everywhere now take the idea of human equality quite seriously. It is no longer accepted as a nature's law that people with darker skins are destined to be servants and those with lighter ones masters. Children of those at the bottom of the social ladder no longer acquiesce in being placed automatically in a similar position. Everybody is entitled to equality. But what is equality? On the authority of the Declaration of Independence, it is a self-evident truth "that all men are created equal." Yet we hear that biology and genetics have demonstrated conclusively that men are unequal. Do biology and genetics really contradict what the Declaration of Independence holds to be a self-evident truth? Or are the words "equal" and "unequal" being used in different senses? Just what have biology and genetics discovered that is relevant to the problem of equality and inequality?

Two geometric figures are said to be equal if they are alike in size and shape and coincide when superimposed on each other. Human equality, whether of persons or of groups, obviously means nothing of this sort. We generally have no difficulty distinguishing between persons whom we meet; the similarity of so-called identical twins strikes us as something unusual and remarkable. In fact, no two persons, not even identical twins, are ever truly identical.

Every human being is in some respect unlike any other. This is not something that modern biology has recently found, but a matter of simple observation so amply documented by the experience of everyone that its validity can hardly be questioned. Every person is, indeed, an individual, unique and unrepeatable. However, diversity and unlikeness, whether of individuals or of groups such as races, should not be confused with inequality. Nor shall the affirmation of equality be taken to imply identity or uniformity.

Biology and genetics have, nevertheless, some relevance to the problem of human equality. They have ushered in a new understanding of the nature and causes of human diversity. This nature resides in the remarkable chemical substances, the deoxyribonucleic acids, which are the principal constituents of the genes carried in the chromosomes of cell nuclei. The genes are products of the evolutionary history of the human species, of the changes which were taking place in response to the environments in which our ancestors were living. The much more recent environments, the living conditions, upbringing, and education of persons and groups also contribute to the diversity which we observe in mankind.

At this point one is tempted to discuss technicalities which, although interesting and even fascinating in their own right, are not absolutely indispensable for our main theme. I shall resist the temptation. The essentials, developed from the seminal discoveries made by Gregor Mendel almost a century ago, can be stated quite simply. How many kinds of genes we inherit from our ancestors is not known with precision; there are at least thousands, and there may be tens of thousands, of genes carried in each human sex cell. An individual carries, then, thousands or tens of thousands of pairs of genes—one member of each pair being of maternal and the other of paternal origin.

What happens when the individual comes, in its turn, to form sex cells, egg cells, or spermatozoa? It follows from Mendel's findings that each sex cell receives one or the other, the maternal or the paternal, member of every gene pair, but not a mixture of the two. A sex cell which receives a maternal copy of a gene A may, however, receive either the maternal or the paternal copy of the gene B, of the gene C, etc. If we take 1000 as a minimum estimate of the number of genes, it turns out that an individual is potentially capable of forming 2^{1000} kinds of sex cells containing different assortments of the maternal and paternal (or grandmaternal and grandpaternal) genes. I say potentially because 2^{1000} is a number far greater than that of the atoms in the universe. Even if only one tenth, i.e., 100, of the maternal and paternal genes are different, the potentially possible number of kinds of sex cells is 2^{100}, which is a 31-place figure, vastly greater than the number of the sex cells produced.

All this adds up to something pretty simple after all: Not even brothers and sisters, children of the same parents, are at all likely to have the same genes. No matter how many people may be born, despite any possible "population explosion," a tremendous majority of the potentially possible human nature will never be realized. A carping critic may remark that we hardly needed to learn genetics to discover what we know from everyday observation, that no two persons are ever alike. Genetics does, however, demonstrate something less commonplace when applied to analysis of the racial diversity among men.

Since our neighbors and even our closest relatives are all distinguishably different persons, it is not in the least surprising that people living in remote lands, or people whose ancestors came from such lands, are more noticeably different. We are inclined, however, to treat the diversity of groups in a manner rather different from diversity of individuals. If asked to describe your brother, or a cousin, or a fellow next door, you will probably say that he is somewhat taller (or shorter), darker (or lighter), heavier (or slimmer) than you, and may add that he is inclined to be kind or easily angered, lazy or impatient, etc. A person whose ancestors lived in America before Columbus is, however, likely to be referred to as an Indian, and one whose ancestors came from tropical Africa, as a Negro.

Up to a point, this is, of course, legitimate. People of African ancestry usually have such conspicuous traits as a dark skin, kinky hair, broad nose, full lips, etc. One should not, however, forget that individual Indians, or Negroes, differ among themselves as much as do persons of the white or any other race or group. When a group of people is given a name, a stereotype is likely to be invented, and, oddly enough, the fewer persons of a given group one knows, the more rigid are the stereotypes of what all Indians, or Negroes, or Irishmen, or Jews are supposed to be. Most unreasonable of all, persons are then likely to be treated not according to what they are as individuals but according the the stereotype of the group to which they belong. This is as unwarrantable biologically as it is ethically iniquitous.

Biologists have found between a million and two million species of animals and plants. An individual animal belongs to a certain species. It is, for example, either a horse *(Equus caballus)*, or an ass *(Equus asinus)* , or a species hybrid (mule), but it cannot belong to two species at the same time. When anthropologists and biologists started to describe and classify races of men and animals, they treated races the same way as they treated species. Each was catalogued and given a name. Again, this was legitimate up to a point. But then difficulties arose. Biologists are sometimes in doubt as to whether certain forms should be regarded as belonging to the same or to two different species; however, with enough material and given careful study, the doubts can usually be resolved. There is, however, no agreement among anthropologists concerning how many races there are in the human species. Opinions vary from two or three to more than two hundred. And this is not a matter of insufficient data; the more studies are made on human populations, the less clear-cut the races become.

The difficulty is fundamental. Biological species are genetically closed systems, races are genetically open ones. Species do not interbreed and do not exchange genes, or do so rarely; they are reproductively isolated. For example, horse genes do not diffuse into the gene pool of the species

Equus asinus, nor do genes of the latter species diffuse into the horse gene pool. Although many mules are produced, they are sterile and do not constitute a channel for a gene exchange. The gene pools of the species man, chimpanzee, gorilla, and orang are quite separate; it is unknown whether viable hybrids between them could be produced by some artificial means. The biological meaning of this separation is the evolutionary independence. Two species may have arisen from a common ancestor who lived long ago, but they embarked on separate evolutionary courses when they became species. No matter how favorable a new gene arising by mutation in the human species may be, it will not benefit the species chimpanzee or vice versa.

Not so with races. Mankind, the human species, was a single evolutionary unit at least since the mid-Pleistocene (the Ice Age). It continues to be a single unit, all segregations and apartheids notwithstanding. Wherever different human populations are sympatric, i.e., geographically intermingled in a common territory as castes or as religious or linguistic groups, some miscegenation and gene exchange crops up, either openly or surreptitiously. More important still is the interbreeding and gene flow among populations of neighboring territories. It is a relative innovation in mankind, that some racially distinct populations live sympatrically, like Negroes and whites over a considerable part of the United States. Before, say, 2000 BC, the human races, like races of most animal species, were largely allopatric, living in different territories. However, the peripheral gene flow, the gene exchange between allopatric but neighboring populations, whether human or animal, is and always was a regular occurrence.

This continuous, sometimes slow but unfailing, gene flow between neighboring clans, tribes, nations, and races upholds the biological and evolutionary unity of mankind. There may be no recorded case of a marriage of an Eskimo with, say, a Melanesian or a Bushman, but there are genetic links between all these populations via the geographically intervening groups. In contrast with distinct species, beneficial genetic change arising in any population anywhere in the world may become a part of the common biological endownment of all races of mankind. This genetic oneness of mankind has been growing steadily since the development of material cultures has made travel and communication between the inhabitants of different countries progressively more and more rapid and easy. What should be stressed however, is that mankind has not become a meaningful biological entity just recently, since men began to travel often and far. The human species was such an entity ever since it became human.

Races, on the contrary, are not, and never were, groups clearly defined biologically. The gene flow between human populations makes race boundaries always more or less blurred. Consider three groups of people, for example, Scandinavians, Japanese, and Congolese. Every individual will probably be easily placeable in one of the three races—white, Mongoloid, Negroid. It will, however, be far from easy to delimit these races if one observes also the inhabitants of the countries geographically intermediate between Scandinavia, Japan, and Congo, respectively. Intermediate countries have intermediate populations, or populations which differ in some characteristics from all previously outlined races. One may try to get out of the difficulty by recognizing several intermediate races; or else, one may speculate that the races were nicely distinct at some time in the past, and got mixed up lately owing to miscegenation. This helps not at all. The more races one sets up, the fuzzier their boundaries become. And the difficulty is by no means confined to man; it occurs as well in many biological species where it cannot be blamed on recent miscegenation.

Much needed light on the nature of races came from studies on the genetics of chemical constituents of the human blood. As far back as 1900, Landsteiner discovered four blood types, or blood groups, distinguishable by simple laboratory tests. These blood groups called O, A, B, and AB, are inherited very simply, according to Mendel's laws. Brothers and sisters, and parents and children may, and quite often do, differ in blood types. An enormous number of investigations have been made, especially in recent years, on the distribution of the blood types among peoples in various parts of the world. Any race or population can be described in terms of the percentages of the four blood types. Almost everywhere persons of all four types are found, but in different proportions. Thus, B and AB bloods are most common among peoples of central Asia and India, A bloods in western Europe and in some American Indians, while many American Indian tribes belong predominantly or even exclusively to the group O.

Several other blood group systems have been discovered, including the Rhesus system which causes some difficulties to the so-called "incompatible" married couples. The genes for these blood groups behave, in general, like those for the "classical" O-A-B-AB blood types. For example, one of the variants (alleles) of the Rhesus gene occurs much more often in the populations of Africa than elsewhere. But mark this well—this gene does occur, albeit infrequently, in human populations almost everywhere in the world, and it is quite certain that it has not spread so widely owing to a Negro admixture in recent centuries.

These facts are profoundly significant. Consider the following situation, which is by no means unusual: A person of European origin, say an Englishman or a Frenchman, has O-type blood, while his brother has B blood. In this particular respect, these brothers differ from each other; one of them resembles many American Indians with O bloods, and the other matches numerous persons in Asia and elsewhere who have B bloods. Or else, one of the brothers may have the kind of Rhesus blood type characteristic of the Africans, and the other brother may have a blood more like a majority of his European neighbors. Such characteristics, of which the persons concerned are usually quite unaware, may become vitally important in some circumstances. If an individual with O-type blood needs a blood transfusion, O-type blood from a donor of no matter what race will be safe, while the blood of the recipient's brother may be dangerous if that brother has A, B, or AB blood.

Every race includes persons with diverse genetic endowments. Genetic studies show that race differences are compounded of the same kinds of genetic elements in which individuals within a race also differ. An individual must always be judged according to what he is, not according to the place of origin of his ancestors. Races may be defined as populations which differ in frequencies, or in prevalence, of some genes. Race differences are relative, not absolute.

This modern race concept, based on findings of genetics, appears to differ from the traditional view so much that it has provoked some misunderstanding and opposition. The use of traits like the blood groups to elucidate the nature of the races of which mankind is composed may seem a questionable procedure. To distinguish races, should one use rather traits like the skin pigmentation, in place of the blood types? Some blood types can be found almost anywhere in the world; on the other hand, pale skins (other than albino) do not occur among the natives of equatorial Africa or of New Guinea, and black skins are not found among the natives of Europe. This objection is beside the point. The blood types are useful because their genetic nature is relatively simple and well understood. The classification of the human races need not be based on any one trait; the behavior of the blood types helps, however, to understand the behavior of other traits, including the skin color.

Though not different in principle from the blood types, the genetics of skin color and similar traits is considerably more complex. The color difference between Negro and white skin is due to joint action of several genes, each of which by itself makes the skin only a little darker or lighter. Geneticists have studied the inheritance of skin pigmentation for half a century, yet exactly how many genes are involved is still unknown. The skin color is obviously variable among the so-called "white" as well as among the "black" peoples, some individuals being darker and others lighter. If we were able to map the geographic distribution of each separate skin color gene as thoroughly as it has been done for some blood group genes, the race differences would probably be resolved into gene frequency differences. It is fair to say that the studies on blood types and similar traits have, thus far at least, helped more to understand races than to classify them.

Another misunderstanding is, in a sense, the converse of the one just considered. Human populations which inhabit different countries differ more often in relative frequencies of genetically simple traits than in any single trait being present in all individuals of one population and always absent in another population. Not only are the differences thus relative rather than absolute, but, to make things more complex still, the variations of different characters are often independent or at least not strongly correlated. Some populations may be clearly different in a gene A but rather similar in a gene B, while other populations may be different in B but less so in A. This makes the drawing of any lines separating different races a rather arbitrary procedure and results in the notorious inability of anthropologists to agree on any race classification yet proposed. Some authors are resolved to cut the Gordian Knot–mankind has no races!

At a risk of multiplying metaphors unduly, this is like throwing out the baby with the bathwater! Race classifiers might have indeed preferred to find simple and tidy races, in which every person would show just the characteristics that his race is supposed to possess. Nature has not been obliging enough to make the races conform to this prescription. Exactly the same difficulties which a student of races encounters in the human species are met with also by zoologists who are working with species of nonhuman animals, and arguments have actually been put forward that animals have no races, or at least that the races should not be described and named. Where, then, are races found? And if they are nowhere to be found, then what are these visibly different populations of the human species which are usually referred to as races?

No more convincing is the argument that the word "race" has been so badly misused by various hate mongers that it became associated in the public mind with unscientific notions and prejudices; let us throw it overboard, and say that mankind has no races, it has only ethnic groups. To me at least, this fails to make sense. If the meaning of "race" is widely misunderstood, as it undoubtedly is, the obligation of a scientist is to set the matter straight, not to resort to subterfuges. A scientist who denies the existence of human races may, regardless of how commendable his motives, actually be helping those whose evil influence he wishes to counteract. Anybody with a pair of serviceable eyes can see the differences between the groups of people which are called races, and a scientist who denies this merely impairs his credibility.

To sum up, the races of man are not integrated biological entities of the sort biological species are. Race boundaries are blurred by the more or less slow but long-sustained gene exchange. The number of races that should be recognized is arbitrary in the sense that it is a matter of convention and convenience whether one should give names to only a few "major" or also to a larger number of "minor" races. An anthropologist who maintains that there are exactly five, or any other fixed number of races is nurturing illusions. On the other hand, there need be nothing arbitrary about race differences; human populations are racially distinct if they differ in the frequencies of some genes, and not distinct if they do not so differ. The presence of race differences can be ascertained, and if they are present, their magnitude can be measured.

The problem that inevitably arises in any discussion of individual and race equality is how consequential the differences among humans really are. Man's bodily structures do not differentiate him very strikingly from other living creatures; it is the psychic, intellectual, or spiritual side of human nature that is truly distinctive of man. Physical race differences supply only the externally visible marks by which the geographic origin of people, or rather of their ancestors, can be identified. The blood types, nose shapes, and skin colors of people whom we meet are so much less important to us than their dispositions, intelligence, and rectitude. It is a person's personality that matters.

The diversity of personalities would seem to be as great, and surely more telling, than the diversity of skin colors or other physical traits. And, though the biological basis of both kinds of diversity is the same in principle, it is different enough in its outward manifestations so that the difference constitutes a genuine problem. This is the perennial nature–nurture problem. The confusion and polemics with which it was beset for a long time were due, in part, to the problem having been wrongly stated—which human traits are due to heredity and which to environment. No trait can arise unless the heredity of the organism makes it possible, and no heredity operates outside of environments. A meaningful way is to ask what part of the diversity observed in a given population is conditioned by the genetic differences between persons composing this population, and what part is due to the upbringing, education, and other environmental variables. Furthermore, the issue must be investigated and solved separately for each function, trait, or characteristic that comes under consideration. Suppose one collects good data on the genetic and environmental components of the observed diversity in the intelligence quotients, or in the resistance to tuberculosis. This would not tell us anything about the diversity of temperaments or about resistance to cancer.

Even correctly stated, the nature–nurture problem remains a formidable one. Dogmatic statements abound on both the hereditarian and the environmentalist side of the controversy, and they usually say much about their authors but not much about the subject at issue. The plain truth is that it is not known just how influential are the genetic variables in psychic or personality traits, or how plastic these traits might be in different environments that can be contrived by modern technology, medicine, and educational methods. There is no way in practice to arrange for a large group of people to be brought up under controlled and uniform conditions in order to see how similar or different they would develop. The converse experiment, observing identical twins, individuals with similar heredities, brought up in different environments, is possible; however, opportunities for such observations are scarcer than one would wish they were.

Some partisans of human equality got themselves in the untenable position of arguing that mankind is genetically uniform with respect to intelligence, ability, and other psychic traits. Actually it is, I think, fair to say that whenever any variable trait in man was at all adequately studied genetically, evidence was found of at least some, though perhaps slight, contribution of genetic differences. Being equal has to be compatible with being different, and different in characters that are relevant to the specifically human estate, not alone in "skin-deep" traits like skin color.

The current civil rights movement in the United States has elicited a rash of racist pamphlets which pretend to prove, very "scientifically" of course, that races cannot be equal because they differ in the average brain size, the average IQ, etc. Now, there is no reason to believe that small differences in the brain volumes are any indication of different mental capacities; the IQ tests are not reliable when administered to people of different sociocultural backgrounds, and in any case they cannot be taken as anything approaching a measurement of the human worth. Be all that as it may, the striking fact, which not even the racists can conceal, is that the race differences in the averages are much smaller than the variations within any race. In other words, large brains and high IQ's of persons of every race are much larger and higher than the averages for their own or any other race. And conversely, the low variants in every race are much below the average for any race. This is a situation quite analogous to what is known about race differences in such traits as blood groups, and in perfect accord with theoretical expectations in populations which exchange genes. Men must be dealt with primarily on the basis of their humanity, and also on the basis of their potentialities and accomplishments as individuals; the practice of treating them according to their race or color is a nefarious one.

It is evidently impossible in an article such as the present one to summarize and to evaluate critically the abundant but often unreliable and contradictory data on the nature–nurture situation in man. It is more useful to consider here some fundamentals that must be kept in mind in dealing with such data. An all too often forgotten and yet most basic fact is that the genes determine not traits or characters but the ways in which the organism responds to the environments. One does not inherit the skin color and intelligence, only genes which make the development of certain colors and intelligence possible. To state the same thing with a slightly different emphasis, we may say that the gene complement determines the path which the development of a person will take, given the sequence of the environments which this person encounters in the process of living. Any developmental process, whether physiological or psychological, can be influenced or modified by genetic as well as by environmental variables. The realization of heredity is manageable, within limits, by physiological and social engineering. What the limits are depends upon our understanding of the developmental processes involved. Modern medicine is able to control the manifestations of some hereditary diseases which not so long ago were incurable. This does not make hereditary defects and diseases harmless or unimportant; even if they can be cured, it is better for the individual and for his society to have no necessity of being cured. This does give substance to the hope that heredity need not be destiny, only conditioning.

Although the mode of inheritance of physical and psychic traits in man is fundamentally the same, their developmental plasticity, the ability to respond to modifying influences of the environment, is different. There is no known way to alter the blood group with which a person is born; it is possible to modify one's skin color, making it somewhat darker or lighter by sun tanning or by lack of sun exposure; the development of personality traits is very much dependent on the family and social environments in which an individual is brought up and lives. The great lability of psychic traits is the reason, or at least one of the reasons, why it is so hard not only to measure precisely the role played by heredity in their variations but even to prove unambiguously that some of these traits are influenced by heredity at all. The more environmentally labile is a trait, the more critical it is for its investigation to have the environment under control; and this is difficult or impossible to achieve with man.

The great environmental plasticity of psychic traits in man is no biological accident. It is an important, even crucial, evolutionary adaptation which distinguishes man from other creatures, including those nearest to him in the zoological system. It is by brain, not by brawn, that man controls his environment. Mankind's singular, and singularly powerful, adaptive instrument is culture. Culture is not inherited through genes, it is acquired by learning from other human beings. The ability to learn, and thus to acquire a culture and to become a member of a society is, however, given by the genetic endowment that is mankind's distinctive biological attribute. In a sense, human genes have surrendered their primacy in human evolution to an entirely new, nonbiological, or superorganic agent, culture. However, it should not be forgotten that this agent is entirely dependent on the human genotype.

A pseudo-biological fallacy, dangerous because it is superficially so plausible, alleges that the differences in psychic traits among human individuals and races are genetically fixed to about the same extent as they are among races or breeds of domestic animals. This overlooks the fact that the behavior of a breed of horses or of dogs is always a part of a complex of characters that are deliberately selected by the breeders to fit the animals for their intended use. A hunting dog with a temperament of a Pekingese, a great Dane behaving like a fox terrier, a draft horse as high-strung as a race horse or vice versa, all these monstrosities would be worthless or even dangerous to their human masters. Man has seen to it that in his domestic animals the genes that stabilize the desirable behavior traits be fixed, and the genes that predispose for variable or undesirable behavior be eliminated.

What is biologically as well as sociologically requisite in man is the exact opposite—to be able to learn whatever mode of behavior fits a job to be done, the mores of the group of which one happens to be a member, a conduct befitting the circumstances and opportunities. Man's paramount adaptive trait is his educability. The biological evolution of mankind has accordingly so shaped the human genotype that an educability is a universal property of all nonpathological individuals. It is a diagnostic character of mankind as a species, not of only some of its races. This universality is no accident either. In all cultures, primitive or advanced, the vital ability is to be able to learn whatever is necessary to become a competent member of some group or society. In advanced civilizations, the variety of function has grown so enormously that learning came to occupy a considerable fraction of the life span. Even where, as in India, the society was splintered for centuries into castes specialized for different occupations, the ability to learn new professions or trades has been preserved.

Champions of human equality have traditionally been environmentalists, conspicuously distrustful of genetic determinisms. Historically, their attitude has been useful in counterbalancing the influence of those racist hereditarians who tried to justify the denial of equality of opportunity to most people on the pretext that the latter are genetically inferior. The environmentalists, however, went too far in their protest. They simply failed to understand that to be equal is not the same thing as to be alike. Equality is a sociological, not a biological ideal. A society may grant equality to its citizens but it cannot make them alike. And what is more, in a society composed of genetically identical individuals, equality would be meaningless; individuals would have to be assigned for different occupations by drawing lots or in some other arbitrary manner. The ideal of equality of opportunity is precious because it holds out a hope that persons and groups diverse in their endowments may enjoy a feeling of belonging and of partnership, and may work for the common good in whatever capacity without loss of their human dignity.

The genetic diversity is a blessing, not a curse. Any society, and particularly any civilized society, has a multitude of diverse vocations and callings to be filled, and new ones are constantly emerging. The human genetically secured educability enables most individuals of all races to be trained for most occupations. This is certainly the basic and fundamental adaptive equality of all mankind, yet this is in no way incompatible with a genetically conditioned diversity of preferences and special abilities. Music is an obnoxious noise to some, ecstatic pleasure to others. Some have a bodily frame that can be trained for championship in wrestling, or running, or sprinting, or weight lifting. Some can develop phenomenal abilities for chess playing, or painting, or composing poetry. Can anybody develop a skill in any of these occupations if he makes sufficient effort? Possibly many people could, to some extent. The point is, however, that what comes easily to some requires great exertion from others, and even then the accomplishment is mediocre at best. The willingness to strive derives, however, at least in part, from a feeling that the labor is rewarded by the thrill of accomplishment or in some other way. There is little stimulus to exert oneself if the results of the exertions are likely to be pitifully small. And it is also possible that there is such a thing as a predisposition to striving and effort.

It is a perversion of the ethic of equality to endeavor to reduce everybody to a uniform level of achievement. "From each according to his ability" is the famous motto of Marxian socialism, and it behooves democracy to grant no less recognition to the diversity of human individualities.

This is not an apology for "rugged individualism"; the "ruggedness" amounts often to indifference or even contempt for individualities of others. Equality is, however, not an end in itself but a means to an end, which can only be the self-actualization of human individuals and the fullest possible realization of their socially valuable capacities and potentialities. Individuals and groups will arrange their lives differently, in accordance with their diverse notions of what form of happiness they wish to pursue. Their contributions to mankind's store of achievements will be different in kind and different in magnitude. The point is, however, that everybody should be able to contribute up to the limit of his ability. To deny the equality of opportunity to persons or groups is evil because this results in wastage of talent, ability, and aptitude, besides being contrary to the basic ethic of humanity.

Limits of Scientific Enquiry

Ernan McMullin

***Ernan McMullin**, born in Donegal, Ireland, received degrees in physics and in theology at Maynooth College; afterwards he pursued courses in theoretical physics at the Dublin Institute of Advanced Studies and later obtained his Ph.D. in philosophy at the Institut Supérieur de Philosophie at Louvain. He has been on the faculty of the University of Notre Dame since 1954. In 1957-1959, he spent two years at Yale on an N.S.F. grant in philosophy of science. In 1964-1965 he was visiting professor in the Department of Philosophy of the University of Minnesota. Since 1965 he has been chairman of the Department of Philosophy at Notre Dame. He is President of the American Catholic Philosophic Association (1966-1967), and member of the Advisory Panel on History and Philosophy of Science and the National Science Foundation (1963-1965). He is a member of Sigma Xi, the Metaphysical Society of America, the Philosophy of Science Association, and the American Philosophical Association. Among his publications are numerous articles on the philosophy and history of science; he translated "Contemporary European Thought and Christian Faith" by A. Dondeyne; he is the editor of "The Concept of Matter" (Notre Dame: University Press) and of "Galileo Galilei" (in press); and he is general editor of the "Fundamentals of Logic" series of Prentice-Hall.*

Limits of Scientific Enquiry

TO SOUND A NOTE of doubt, of limit, in a series of lectures commemorating man's most recent conquests of nature is an unenviable task. It is one that does not come easily to me, since I am much more familiar with the opposite role, that of pleading the glory of science to humanists who are antagonized and overwhelmed by its power. I am speaking here mainly to scientists, to scientists whose implicit metaphor as they speculate about the future of their science is the boundless reach of space-time itself, into which we have begun to take our first nervous but exhilarating steps. Before our rockets and our telescopes lies a universe that a billion years of earthman's exploration would not exhaust. It is tempting to assume that our science, the instrument that has brought such soaring exploration into the reach of our blueprints, has an equally illimitable future. To suggest that such may by no means be the case has almost an impiety about it in these heady days of successive dramatic "breakthroughs." I come then to my topic with diffidence and a certain amount of trepidation, somewhat as an agnostic might approach the opportunity to speak at a Revivalist meeting.

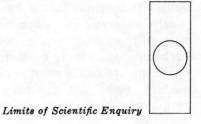

You notice that I said "agnostic," not "atheist." For the most part, I shall content myself with raising certain doubts and recalling some overlooked historical analogies. I am not about to claim that scientific enquiry is nearing its end, nor even to maintain that it must have a term at some finite time in the future. My aim is a modest one really: to argue that some limits on the future progress of scientific enquiry are possible, and certain among them probable. The aim is modest, yet if it be achieved, it would challenge one of the most pervasive convictions of twentieth century man: that the progress of science is likely to remain the steady and inevitable affair it seems recently to have become.

If someone had been asked in 1865 to forecast what the next century would bring by way of scientific achievement, his prediction would almost certainly have been very wide of the truth. How can one answer a question of this sort at all? It is certainly not an ordinary scientific question, since it does not lend itself to answer in terms of the familiar theory-construction and testing of the scientist. Ought we conclude, then, that no answer is possible, and that the search for one is therefore an idle whim? Certainly no definitive answer is possible, but surely there are some sources of evidence available that make a tentative limited prediction conceivable. These sources are of three sorts. The principal one is the history of science itself. A careful look at its past, especially its recent past, may give some hints as to the shapes its future is (or is not) likely to take. It is true that "scientific revolutions" occur; yet they do not altogether obliterate familiar landmarks, nor are they entirely without warning or analogy. The second source is the philosophy of science, the philosophical analysis of method and concept and implication. There is a sense in which a philosophical moment has been central to every scientific change that can properly be called "revolution"; the sort of analysis that led a Newton or an Einstein to a new plateau was at least as much philosophical as scientific in character. If one is to search the more distant future, it is not to the routine expressions of respectable scientific theories following accepted "scientific method" that one would first look; one would ask rather about the "revolutions" of the future, what would trigger them, what direction they might take. The third source of evidence is the contemporary theories of science itself. Are there any indications in them of shoals ahead, of barriers that cannot be crossed?

It is worth emphasis that the contemporary view of scientific enquiry as having an illimitable horizon is of much more recent origin than is often supposed. We have lived with this idea in all its dramatic force so intimately that it is easy for us to forget that very few people, in or out of science, would have made any such prognosis before the present century. It will be instructive to look closely at this earlier phase of science in order to understand better the later course of events.

PART ONE: SCIENCE AS A TERMINABLE ENTERPRISE

§1. *The Greek Ideal of Science*

There were two factors that led Greek philosophers to regard science (*epistémé*) as a finitary affair all through those "golden centuries" between 500 BC and 300 BC, when the first groundwork of natural science was being laid. First was the close association in the Greek mind of perfection and of limit. A thing is what it is, is known for what it is, in terms of its boundary. A perfection is fully possessed by (or a predicate is unqualifiedly predicable of) a subject if the definition of that perfection is fully realized in the activity of the subject. Definitions are in principle always possible, and a good definition is "definite," i.e., marks off the "finis" or boundary within which the concept applies unequivocally. In this perspective, unlimit connotes imperfection, a lack of definiteness, of knowability, and ultimately, of being. The only unlimited (or infinite) entity in Greek thought is primary matter, lowest in the scale of being, a "being" only by an extended use of the term. The ultimate source of true being in Plato's universe is the Forms; they themselves are in a sense constituted by their very finiteness, their relations with one another, and with the most perfectly bounded of all Forms, the Form of One. Aristotle's First Mover is Pure Actuality, but the overtone of "pure" or "unlimited" here conveys lack of potentiality, no more. If there was one thing that Aristotle was relatively sure of, it was that the notion of an actual infinite made no sense.

In such a context, to suggest that the process of enquiry might have no end, would imply that knowledge is necessarily imperfect, and that "true" knowledge (*epistémé*) is thus unattainable. In a civilization so intent on knowledge, so impressed by the first successes of speculative and predictive thinking, such an admission could hardly have been made. True, Plato seems to have realized in his later years how difficult of attainment *epistémé* is; in the *Phaedo*, he resigns himself to a hypothetical approach to the all-important question of the nature of soul, and suggests with great insight how this approach can be made to work. And Aristotle in his biological works found the standards of *epistémé* notoriously hard to live up to. Yet neither of these great thinkers would ever have yielded up their conviction that *epistémé* of a definitive sort is possible, even though we may in practice frequently have to be content with *doxa*, true opinion.

The main reason for their conviction was the existence of the newly discovered science of deductive geometry, which gave a pattern of what *epistémé* could—and presumably should—be. There appeared to be a finality, a necessity, about the axioms of geometry, and the theorems derived from them. The only way (it seemed) in which geometry could develop further was by addition, the addition of new theorems and possibly new axioms for more specialized purposes. Science consists of certain truths which are admittedly often hard to come by, but once "seen," carry their own warrant with them. They are capable in principle of being fully grasped; indeed, until they are fully grasped, one cannot have the sort of assurance about them that is proper to *epistémé*. They are not reformable nor do they need development. Even in domains of natural science, like biology, where such principles are extremely difficult to obtain, the hope was to uncover an intrinsic structure of intelligible relations between living species, using definitions by genus and property. If the basic concepts and the relations could once be defined, biology could become as definitive a science as geometry.[1]

But to pose the issue in this way is perhaps misleading, because it might suggest that the Greek philosophers were explicitly concerned with this question of whether or not *epistémé* is final in its declarations. It is easy to say how they would have answered the question. That they rarely posed it is, however, due to another factor, the last one we have to take into account in exploring the genesis of the Greek view of science. The notion of historical development was altogether foreign to their world-view. Change—yes; biological growth—yes. But a cumulative and significant change that does not simply terminate (as the growth of a biological individual does) in an instance of the "unchanging" species, this was something they were not prepared for. History either has no overall pattern, and is simply a series of human gropings around an unchanging set of themes, or else it is cyclic, so that decay is just as inevitable as is development, and thus both are without ultimate significance. Knowledge may be cumulative, but it has been gained and lost countless times before. And it does not have within it a dynamism of discovery that would lead one generation to build on the researches of the previous one. Indeed, the notion of research itself (in the sense familiar to us today) is not quite at home here.

True, the astronomy of Eudoxus was built upon the labors of generations of patient observers of planetary positions. And the medical treatises of the Hippocratic school relied heavily upon a continuous tradition of controlled enquiry in domains like anatomy and physiology. But these were at most cumulative, not really developmental. The hope of

[1] This point is discussed by many of the essayists in: Aristote et méthode (Louvain: Nauwelaerts, 1961), see also E. McMullin, "The Nature of Scientific Knowledge: What Makes It Science?" in: Philosophy in a Technological Culture, ed. by G. McLean (Washington: Catholic University Press, 1964), pp. 28-54.

the Greek astronomer was to find a formula that would express the path of his planet in a definitive and elegant mathematical figure. And though his science (like that of the medical men) did contain the methodological hints that might have led to a very different ideal of "true science," these hints were not adverted to, nor could they have been in that springtime of speculative knowledge, when the boundaries still seemed so near and so clear.

§ 2. *Slow Change*

The decline of natural science that began in the Hellenistic period of the popular textbook was not reversed until fifteen hundred years later with the introduction of Aristotle's forgotten "natural" works into the West around 1200 AD. But in the meantime, the spread of Christianity had profoundly modified many of the assumptions and the concepts that had gone to the shaping of the Greek ideal of a non-historical, intuitively warranted *epistémé*. That ideal itself would be much longer in changing, but most of the ingredients of change were already present by 1300 AD.

First, there was a new vision of history in which time was no longer the endless returning wheel; it now had a significance in itself. The story of God's dealings with the Chosen People was set in time; it had a beginning; it looked to an end; everything that happened was significant. Above all, Christ's entry into time was once and for all; in Him, God had transformed human history. The most profound of truths was thus no geometrical *epistémé*, but a vision of history as leading up to and somehow becoming one with the Son of God. Augustine cast theology in the role of the supreme science, but he knew its dependence on the language of men too well to attribute to it the unchanging sure grasp that is possible in mathematics. True science is no longer simply of the unchanging; but there is still no clear notion of it as developing.

The old link between finitude and perfection had also been broken. Plotinus and Aquinas between them had subtly redefined the entire set of metaphysical concepts so that limitation is seen as coming from an ontological lack of some sort, and God is now thought of as infinite. The sharp edges of the Platonic Forms dissolve in the analogy of Aquinas' crucial concept of being. The cosmology of the day still imposed tight spheres around the earth, but there was a growing tension between the closed universe of the scientist and the infinite reaches of the metaphysician. The impulse to burst through the spheres of Aristotle came in these centuries from theology or metaphysics rather than from physics or astronomy—and from theology perhaps even more than metaphysics.[2]

One of the central doctrines in Christian theology is that of the freedom of God. Freedom is one of those defining attributes that allow man to be called the "image of God." Obviously, God's freedom will differ very much from man's; his power is unlimited, so there are none of those obstacles or limits that go to the shaping of human freedom. Nevertheless, Christian theologians from the beginning insisted that the work of Creation must be regarded as in some basic sense free. The universe and its history might have been otherwise; if man can by his choice of one alternative over another make a difference to history then so, in a transcendent sense, can God. There is clearly a tension between this view and the rationalistic strain that we have already seen in Greek thought. If *epistémé* is composed of truths that are necessary, that carry their own intrinsic warrant once the concepts expressing them be properly understood, then *epistémé* cannot be other than it is. It cannot depend (so it would seem) upon a free choice, not even upon God's choice. If physics is to be an *epistémé*, as Aristotle had hoped it would be, its necessary character would appear to suggest that the physical universe in its general features (those tested in Aristotelian physics) could not be other than it is.

[2] See E. McMullin, "Four Senses of Potency" in: The Concept of Matter (Notre Dame: University of Notre Dame Press, 1963), pp. 295-318, esp. pp. 298-301.

The gradual introduction of Aristotle's "Physics" into the universities of Western Europe between 1200 and 1260 brought this problem to the fore. The university theologians, most of whom were Augustinian in inspiration, were uneasy about many features of the Aristotelian enterprise, not least about the self-sufficiency it seemed implicitly to attribute to the human intellect. Philosophy to them was a subalternate science, needing the illumination of theology; a true *epistémé* of Nature (in the strict Greek sense of that term) unduly diminished the primacy of theological truth. Their opposition to the Averroist emphasis on the autonomy of philosophy led to the famous condemnation of various Averroist theses by a council of French bishops at Sens in 1277. In his great work on the history of science, *Le systeme du monde*, Duhem traced the beginnings of a new, more qualified, view of the capacities of the human intellect to that reaction. If God is truly free in His creation, as some philosophers now begin to suggest, men will have to work harder to find out the details of that creation, and will have to be more tentative in their claims to scientific knowledge about it.

In the century that followed 1277, two movements of thought are worth noting in this connection. One was the growth of empiricism in the "nominalist" school of Paris. Nicholas of Autrecourt maintained that since our knowledge comes to us only through the senses, tentative generalizations from sense-knowledge are the best we can do in the domain of nature. The certainties of mathematics are nonempirical in character and tell us nothing about the world. The nominalists tended to regard the older ideal of a physical *epistémé* as unattainable; in their view, physics had to be a tentative, groping, observation-bound affair. Ockham still allowed some limited demonstrations in physics, but the nominalist theories of causality and their denial of all reality to universals made natural science appear a rather uncertain domain, at best.[3]

[3] See E. Grant, "Late Medieval Thought, Copernicus, and the Scientific Revolution," Journ. Hist. Ideas, 43 (1962), 197-220.

The stress on the primacy of will in speaking of God also increased. Scotus insisted that though God is a necessary being, his relation with finite being is radically contingent. He rejected the Aristotelian linking of First Mover to sublunary effects by a series of necessary causal relations: such a scheme would imply the impossibility of God's intervening in a free singular way in his own Creation. The Scotists and the nominalists differed on most philosophical issues, but they were at one in rejecting the achievable and definitive physical science-by-demonstration proposed by their Aristotelian colleagues.

The effects of this scepticism on natural enquiry were, not surprisingly, catastrophic. By the end of the fourteenth century, little was left of the *élan* of physical research in the golden days of the Aristotelian revival. But one positive effect of all this criticism was the suggestion of a more limited ideal of science itself, one in which experience would play a more decisive role, and in which continuing enquiry would be necessary. The discussions among astronomers about the possibility of a plurality of worlds, the presentation of positions in physics as plausible hypotheses requiring observational confirmation—these were the fourteenth century presage of a gradual underground shift in the theory of science itself, a shift that was to be far more important for the future of science than any specific physical discoveries would have been.[4]

§ 3. *The Seventeenth Century*

During the two and a half centuries that separated Buridan from Galileo, the debate between Thomists, Averroists, Scotists, nominalists, and the rest, rose to a crescendo and then was silenced by the religious crisis of the sixteenth century. Much of what was important in the natural science of those centuries went on outside the universities; people like da Vinci or Copernicus had little to do with the

[4] I have discussed the thesis of this section more fully in "Medieval and Modern Science: Continuity or Discontinuity?" in: Intern. Philos. Quarterly, 5 (1965), 103-129.

natural science of the lecture halls. The voyages of discovery across the oceans gave men the feeling that the horizons of human effort could well lie further off than had been realized. Yet when we look at the programs laid down for science by the great figures of the early seventeenth century, Bacon, Galileo, Descartes, Boyle, we find that they still tend to assume a definitive achievable ideal of science, just as the Greeks had done. Their manner of attaining it would be quite different, of course, but they all felt that it could be attained. Each had his own reasons for believing physical enquiry to be capable of yielding a definitive *epistémé*. Bacon thought he had hit upon a method of discovery that would permit science to progress "as if by machinery."[5] Galileo thought that nature could be exhaustively and exactly described in mathematical formulae, such as his own "law" of falling bodies. Descartes tried for a time to carry through the geometrico-deductive ideal of science outlined in the *Regulae*. Indeed, it might well seem that the caution of fourteenth century empiricists and theologians was forgotten in the excitement of the "new science," and that all the prerogatives of the old were claimed for it.

Yet there were many signs of a more tentative spirit. Descartes was just as anxious to safeguard the freedom of God as the medieval theologians had been; observation is our only way of finding out which of the possible worlds God actually chose to create. In the *Discourse on Method*, he concedes the importance of hypothesis; it is not clear, however, to what extent he would have allowed it a role in the finished product of the scientist. Galileo often speaks like a rationalist, but his skillful selection of hypotheses and corroboration of them by appealing to observation (in the discussion of the nature of the lunar surface in the *Dialogo*, for instance) leads the reader to be wary in making general-

[5] This tension between the two ideals of discovery: individual creativity and rule-governed "machinery," is discussed in my "Freedom, Creativity and Scientific Discovery" in: Freedom and Man, ed. by J,C. Murray (New York; P. J. Kenedy & Sons, 1965), pp. 105-130.

izations about Galilean methodology. Bacon sees unlimited horizons ahead for science, but only in its pragmatic aspect—the aspect, however, that he considers primary. There will (he urges) be no end to the changes man will be able to impose upon nature to his own benefit.

If someone had posed the question: "Is natural science a once for all definitive affair or is it a matter of enquiry?" to a contemporary of Newton's, it is hard to know how he would have answered. By that time, there were two strong arguments for seeing science as a progressive unending quest. The first came from the discoveries recently made in the realms of the very small and the very distant by means of the microscope and the telescope. Popular imagination had been greatly excited by the discoveries of worlds of little creatures in apparently clear drops of water. Pascal wondered whether these worlds might not contain yet more minute worlds and so on indefinitely, a natural question once the first sublevel is established. And extending outwards from man's once-mighty abode, the earth, was an unimaginable gulf with an occasional point of light—remote, mysterious, unreachable. It might have seemed that scientific knowledge had no more a boundary than did the power of microscope or telescope.

Second, an appreciation for the central roles of hypothesis and approximation was growing. In a well-known passage in his *Treatise on Light*, Huyghens suggested a new ideal of scientific evidence:

> "A kind of demonstration which does not carry with it as high a degree of certainty as that employed in geometry, and which differs distinctly from the method employed by geometers who prove their propositions by well-established and incontrovertible principles. Whereas here (in optics) principles are tested by the inferences which are derivable from them. The nature of the subject permits of no other treatment. It is possible, however, in this way to establish a probability which is little short of certainty. This is the case: 1) when the consequences of the assumed principles are in perfect accord with the observed phenomena; 2) more specially, when these verifications are numerous; 3) but above all when one employs the hypothesis to predict new phenomena and finds his expectations realized."[6]

[6] Preface; Crew translation.

This "hypothetico-deductive" ideal of scientific warrant is in sharp contrast with the Aristotelian one it gradually replaced, for it forces the scientist to look outside the theory he is testing for its final justification; it will not suffice just to grasp the conceptual structure of the theory itself and decide whether it is "necessarily true" or not.

An experimental warrant is necessarily a limited warrant, one allowing for future development as experiments become more precise and new parameters are introduced. And to the extent that a science is hypothetical, it allows for the possibility of a counterhypothesis, for a fitting and trying that may well prove endless. The tension between a rationalist and an empiricist ideal of scientific evidence ought (one might suppose) have led to some tension between a finitary and a nonfinitary view of scientific enquiry. The rationalist will define science in terms of certitude and intrinsic evidence; experiment will be seen as an occasion of discovery rather than as evidential support. The empiricist is likely to stress induction and the gradual amassing of data, the sharpening of experience by instrumentation, and so on.

Yet when one looks at the writings of scientists, not only in the seventeenth century but even as late as the nineteenth, it is obvious that these philosophical labels "rationalist," "empiricist," do not fit very neatly. Not only do we find scientists—Galileo and Descartes would be good examples—swinging from "rationalist" to "empiricist" theories of evidence, even in the same book, but also it is quite clear that the finitary view of science as terminable is still dominant even among those who are most empiricist in their emphasis. Huyghens and Newton are well enough aware of the hypothetical nature of much of what they do, yet they still (Newton especially) appear to regard science as something once for all accomplished, section by section. Even Hume (within whose philosophy a finitary notion of science makes little sense) is loth to break with the tradition; a developmental science would have fitted in very well with his inductive approach, yet he does not appear to think of the science of his day in those terms.

This is all the stranger in that this was the century of history, when temporal dimensions were being searched out with a new sensitivity. Explanation in terms of development, of patterned historical change, was gradually entering into geology, botany, and historiography itself. Yet curiously enough, when we look at the works of those who stressed various aspects of process, history, development, men like Herder, Goethe, Hegel, we do not find much stress on science as one of those entities with a history, a definite, continuing, and perhaps unending development. Nature may develop, may evolve (though a full realization of this will not be found until long after Hegel), but the science of nature? Philosophy has had a history and a definite dialectic. But what of mechanics? The great German thinkers of the eighteenth and nineteenth centuries had much less to say on this. And for one reason, principally, it would seem.

§ 4. *The Newtonian Years*

That reason was Newtonian mechanics, the most pervasive intellectual feature of their world. There was something block-like, inevitable, final about it right from the beginning. Newton himself had said that his "laws of motion" were not "hypotheses," but "deduced from phenomena and made general by induction, which is the highest evidence that a proposition can have in this philosophy." They are established by experiment, whereas an hypothesis is assumed without experimental proof, for the purposes of explanation. In optics, Newton was prepared to allow hypotheses, but he warns that even though he sometimes asserts them without qualification, they ought not be confounded with his "other discourses," his mechanics; the certainty of one ought not be measured by the certainty of the other.[8] His editor, Roger Cotes, was even more definite about the matter. In his preface to the second edition of the *Principia*, he distinguished between two sorts of philosophers. First

[7] Letter to Cotes, Newton's Philosophy of Nature, ed. H. S. Thayer (New York; Hafner Publishing Co., Inc. 1953), p. 6.

[8] Letter to Oldenburg.

there are "those who assume hypotheses as first principles of their speculations" and who, though they thereafter "proceed with the greater accuracy," can arrive (because of their starting-point) at no more than an "ingenious romance." Then there are those who "assume no principle not proved by phenomena"; they "deduce from select phenomena... the simpler laws of forces." Newton is preeminent among these latter; others had discussed gravity before him but he "was the only and the first that could demonstrate it from appearances and make it a solid foundation to the most noble speculations."[9] In other words, though some parts of science might be speculative, the Newtonian mechanics is conclusive and basically complete.

To see why this claim was so plausible, let us go back for a moment to Galileo, who was similarly deceived about the nature of mechanical laws. Once he hit upon his law of falling bodies ($s = gt^2/2$, in modern notation), it seemed immediately as though it could not possibly be otherwise. From the very definition of acceleration itself, once it was properly grasped (and grasping it was of course Galileo's great achievement), the law would seem to follow in a purely analytic mathematical way (by the equivalent of simple integration). Was there anything empirical about it at all, then, other perhaps than its mode of discovery Galileo thought it conclusively demonstrated. And it was certain that the varied theorems he proved for uniformly accelerated motion were mathematically true. But, as his dogged Aristotelian "straight man" objected, how could he be so sure that this was in fact "the motion one meets in nature"?[10] In other words, how could one be sure that the motion of free fall of physical bodies *is* uniformly accelerated? Here Galileo conceded that experimental support was necessary, and he described his famous inclined plane. The assumption he had made was that two bodies dropped

[9] *Op. cit.*, pp. 83-84.

[10] Two New Sciences, Crew-de Salvio translation (New York: The Macmillan Company 1914), p. 178.

from the same height along different smooth planes will acquire the same speeds, "a postulate, the absolute truth of which will be established when we find that the inferences from it correspond with . . . experiment."[11]

But, of course, it was not true that the motion of free fall on the earth's surface is uniformly accelerated; as Newton was afterwards to show, the acceleration increases as the body approaches the center of the earth. But to a fair degree of approximation, Galileo's assumption was correct. The "demonstrative" character he attributed to his result was due to a combination of two factors: 1) the demonstrative character of the mathematical analysis by which the theorems derivable from his original definitions could be obtained; 2) the plausible nature of the inductive assumption he had made (on the very slender basis of a few experiments in which the crucial parameter—distance from the center of the earth—was not varied), the assumption that natural acceleration took place in a manner "exceedingly simple and rather obvious to everybody," i.e., uniformly.[12] Galileo realized that with this newly defined property he could fully describe a motion in spatiotemporal terms; there seemed to him, then, to be every reason to suppose that in the case of "natural" motion (free fall), this crucial property would remain constant. So reasonable an assumption was it that when it was roughly confirmed by some experiments, the formula became in his mind not only mathematically derivable but physically true of the world in which we live.

This pattern was to be followed many times in the later history of mechanics. There is something about mechanics that makes it very easy to cast as an *a priori* science, "seen to be necessarily true," once its concepts are fully grasped. These concepts may all too easily be regarded as purely "mathematical," just because the system works in an

[11] *Op. cit.*, p. 172.

[12] *Op. cit.*, p. 161.

analytic-deductive way. And it is possible to see in mechanics an algebraic system, a locus for interesting mathematical problems, as generations of textbooks of "rational mechanics" testify. But, in fact, mechanics is not a part of mathematics, though it is entirely dependent upon mathematical syntax. Its concepts derive their origin and meaning (their semantics) from a complex set of physical operations and physical categories; and mechanics can only be said to be true about the world on the basis of continuing experiential support. The "opening towards rationalism" that mechanics has always displayed, depended upon the assumption that basic physical concepts (*space, time, acceleration. . .*) can be fully defined by an axiomatic system. But if they are to apply to the world, operational procedures have to be specified, and physical hypotheses ("the acceleration of free fall is uniform") have to be confirmed.

This can be very clearly seen when we look at Newton's formulation of mechanics. He cast it in quasi-axiomatic form, and called it: *The Mathematical Principles of Natural Philosophy*. His "laws" appear at first sight to be statements of empirical regularity, but on closer scrutiny seem to be no more than definitions of the abstract concepts of force, mass, and inertial motion. Even when he gets to specific laws of force, such as the inverse-square law of gravitational attraction, was it not plain from a simple geometrical diagram that anything propagated from a center would diminish in intensity outwards according to a law of squares ? In a three-dimensional Euclidean world, propagation of a real entity could not follow any other law. There was no need to worry about it coming out as a 2.0001 power on the basis of more accurate measurement. The integer value was based on geometry, not on measurement.

This kind of rationalist philosophy of science became all the more attractive as the eighteenth century wore on, and Newtonian mechanics went from one success to another. It seemed as though the fundamental laws of physical motion had been discovered, and that they were just what one would expect them to be in a mathematically simple universe. True, there were stubborn areas still under discussion: light, electricity, heat, where alternative "hypotheses" were still open. But then heat gradually lent itself to analysis in purely mechanical-statistical terms. Why should not light and electricity also be resolved as complex mechanical phenomena with their own special "laws of force," but otherwise purely Newtonian in character? After all, the fundamental concepts (*space, time, motion, force, mass*) would have to be the same.

Kant, in his youth an able physicist, gave expression to this rationalist view of physics:

> "When Galileo let balls of a particular weight... roll down an inclined plane... a new light flashed on all students of nature. They comprehended that reason has insight into that only which she herself produces on her own plan, and that she must move forward with the principles of her judgments according to fixed law, and compel nature to answer her questions... Reason, holding in one hand its principles... and in the other hand the experiment which it has devised according to those principles, must approach nature in order to be taught by it: not in the character of a pupil who agrees with the master, but rather as an appointed judge who compels the witnesses to answer the questions which he himself proposes... Thus only has the study of nature entered on the secure method of a science after having for centuries groped in the dark."[13]

Could physics, then, be part of the transcendental philosophy? Could its laws and categories derive their universality and permanence from the minds of those who impose them (via their "questions") on nature? Here Kant made a distinction between "pure" physics, whose judgements are synthetic *a priori*, and "empirical" physics which is *a posteriori*. Newtonian mechanics forms the basis of "pure" physics. Space and time are *a priori* particulars; Euclidean geometry and Newtonian mechanics express their articulations, in a necessary way. Space cannot be perceived as other than Euclidean; the presuppositions, listed by Newton in his *General Scholium* to the *Principia* are necessary for any adequate

[13] Critique of Pure Reason, Preface to the second edn. (1787).

description of motion. Kant made it quite clear that he believed Newton's mechanics to be an ultimate and permanent achievement of the human mind.[14] And most of his contemporaries, even the more empirically-minded Britons across the Channel, would have concurred with him in this.

§ 5. *The Future of Science a Century Ago*

How would a scientist of 1865 have thought of the future of his science? Did it seem illimitable? What was the status of "past" science in the century of progress? Oddly enough, this topic was not often explicitly raised. But one can gather from incidental references that the history of science appeared to most scientists like the opening of one room after another in a large, but not infinitely large, mansion. After each room was fully explored, there was no more to say, and one went on to the next. The basic motions of material bodies were (it seemed) fully understood in terms of mechanics; only the practical impossibility of providing a state-description for enormous numbers of particles prevented Laplace's ideal of complete prediction from being fulfilled. Some processes (electrical, chemical psychological) did not, it was true, themselves seem to lead to explanation in straightforward mechanical terms. But much progress had been made in the program of reduction to mechanics: in that very year of 1865, Maxwell's paper on propagation showed a fundamental affinity between light and electromagnetism, and held out hope of a mechanical theory that would cover both. It would involve new "higher-level" laws, just as chemistry was already known to do, but these laws would simply be superimposed upon the primary Newtonian laws, to cover the more complex forms of process. Newton's laws would still exactly describe the local movements of the physical elements (ether, corpuscles...) within the higher-level science. There was no more to be

[14] See Gottfried Martin, Kant's Metaphysics and Theory of Science (London, 1956).

said about these lowest-level movements than Newton had already said. The mechanics of 1865 was thus completed; a great past, but nothing of interest in its future. Optics seemed to be steadily moving in the same direction, though the ether concept was causing much trouble.

Even in chemistry, despite the vast number of compounds still to investigate, it seemed as though the frontier might not be far off. If Dalton's atoms really were "atoms," i.e., structureless, indivisible entities, all that remained to be discovered were the various stable configurations and the laws of force operating within them: in other words, chemical mechanics. It is worth emphasizing that from the point of view of 1865, the atoms might really have turned out to be "atoms"; there was nothing logically impossible about this. A universe consisting of unstructured indivisible atoms is quite conceivable, it would seem. And the physics of such a universe might quickly reach its term. There would be awkward questions about the interactions of radiation and atoms: for example, could such atoms be colored i.e., would there be some mechanism available to explain the light-emission from (or colors of) material objects? And could they combine chemically in all the various ways that compounds do? There was nothing in the optics or chemistry of 1865 that absolutely excluded such hopes. Had we been in a truly "atomic" universe, had the atoms been those of 1865 chemistry, it is likely that chemistry and physics would before this have virtually reached completion.

In putting the matter thus, it is important not to give the impression that our universe might just as easily have been strictly atomic. This is certainly not true in terms of today's knowledge: today a whole profusion of newly discovered processes forces us to structure the "atom" as our only available means of explanation. But it would not even have been true in terms of the knowledge of 1865 either, though this is less evident. Those who implicitly assumed in 1865 that atoms really were "atoms" (i.e., that the hypothetical model of the chemist was a more or less exact description) had to suppose not only that all the known properties of matter could be explained in terms of their relative motions and configurations (an ideal that even two millennia of familarity with Democritus had never really rendered plausible), but also that all yet-to-be discovered properties could be handled in the same way. In the Newtonian–Daltonian climate, such a hope might have seemed warranted. Yet it was an audacious one, when one reflects on the wide diversity of physical processes and the extremely limited range of the indivisible mechanical atom as an explanatory paradigm.

Let us look back one last time at those scientists of the 1860's and ask what they were doing. The physicists were beginning to work on field theories which unified the three great domains of electricity, magnetism, and light. Maxwell's equations could most easily be interpreted in terms of rapid oscillations of electrical and magnetic forces in the ether. It was not clear how such forces could propagate as mechanical "waves," yet there appeared to be every reason to hope that the two new sorts of force could be added to gravitational attraction, and a general mechanics established for all three. But the new science of spectroscopy raised some nagging questions. At first it seemed as though the lines in the spectrum of each element could be produced by different atomic and molecular vibrations. But when hundreds of lines were discovered (in the spectrum of iron, for instance), it began to look as though the hitherto unstructured "atom" would have to have some sort of internal differentiation.

In chemistry, attention was focused on the various sorts of chemical interaction, and especially on the problem of electrolytes. Faraday had shown as far back as 1833 that electrical charge is carried in uniform unit amounts by ions, the active atomic constituents in electrolysis. Research on ionic exchange continued and met with notable success. But it seemed increasingly difficult, once again, to assume that all the numerous valencies discovered (and already exploited) in organic chemistry were going to be explicable in terms of a mechanical–electrical model of an unstructured atom. And it was also difficult to know how a continuous field theory and a discontinuous charge theory could be unified in a single theory of electricity.

The 1860's were, of course, one of the great decades for biology. Ever since Aristotle, the aim of biologists had been to find some way of linking living beings in a scientific ordered way. Much had been done on classification. But it all seemed rather arbitrary: there was a hierarchy of sorts, a *scala naturae*, but what more could one say? Darwin changed all that. He succeeded in adapting genetic explanation, already fairly successful in geology, to the problem of classification. The idea of evolution was not new; even in biology, it had been talked about for nearly a century. But Darwin introduced a theory of evolution (involving a complex of explanatory elements) that began to bring together the data of comparative anatomy, palaeontology, and the rest. Immediately, a vast domain of history opened to scientific inspection, a domain so complex, so extended, and so contingent in its present-day remains, that the future task of biology seemed almost endless.

In summary, then, after several centuries of growingly intensive development, the natural sciences towards the end of the nineteenth century had attained to powerful and highly successful explanatory schemata. They seemed relatively definitive; in physics and chemistry, at least, the fundamental metaphor, mechanism, was fully grasped. Research in these areas was thus thought as though one were entering one after another, a series of rooms, each of which could be fully explored before going on to the next. Physical theory was not regarded as continuously progressive; in their more sanguine moments, many physicists would have agreed with Michelson's famous comment that it only remained to add further decimal places to measurement. The older Greek view of science, then, as a finitary enterprise was still not far from the surface, though the exhausting of its more complicated areas—biology, psychology—might take a long time.

PART TWO: THE HORIZONS RECEDE

§ 6. *Atomic Substructure*

As the century drew to a close, this ideal began to seem more and more remote. By the 1920's, even physicists were hesitant to claim any sort of finality for their basic theories, and Newtonian mechanism had been recognized as a useful wayside approximation, its *a priori* character shattered irremediably. Of the many factors that contributed to this dramatic change, three can be singled out for special mention.

First came the "discovery" of the electron. Cathode rays were known since 1869, but it took decades of clever and patient interplay between experiment and theory to show (1897) that they had to be regarded as streams of fast-moving electrified particles, whose mass was far smaller than that of an atom. Furthermore, the ions of chemistry now seemed explainable as atoms which had lost a component electron, so that the electron was accepted as in some sense a constituent of the atom. At the same time came the discovery of radioactivity (1896), which soon was seen to give rise to three different sorts of "ray," two of them apparently consisting of particles emanating from the atom. And the effect of this emission was to change the chemical nature of the atom—a series of consecutive emissions can change an atom of uranium into an atom of lead. Rutherford's scattering experiments (1911) showed that the atom had a "nuclear" structure, one tiny massive scattering center. Bohr completed this picture with the placing of the electrons in planetary orbits on the basis of an application of quantum ideas to spectroscopic data.

Thus, the formerly structureless atom is now seen to be a complex of electrons and heavy nuclei. These nuclei will turn out to "contain" other particles: protons and neutrons first, and mesons later. We shall not follow these more recent developments, with their competing models of nuclear structure, still giving rise to considerable disagreement. The important thing was that the notion of the likelihood of substructure was becoming firmly entrenched. Not without a struggle, though. The new "particles" for a time took on the same ultimate ontological character that their predecessors, the atoms, had done. Many scientists, and most nonscientists, spoke of them as unstructured and as ultimate. Eddington, a very great physicist in his own right, even tried to deduce by means of a "transcendental" analysis of the conditions of experimental measurement, the number of these "ultimate" particles (electrons, nucleons) in the universe. His ingenious speculation was, however, rapidly undermined by the discovery of various types of mesons. It became clear that the "ultimates" in a particular physical theory ought not be taken as ultimate in any more significant way. A later theory may well be forced to structure what was at first (for simplicity's sake) postulated as unstructured. Today, dozens of subatomic entities are known, not just two or three, and strenuous efforts are being made to find a theory that will unify this multiplicity, as the Bohr-Rutherford theory did the chemical elements. But no one would want to say today (as so many did of the elements) that the basic entities of the new theory will be unstructured ultimates.

This new realization recalls a philosophical puzzle as old as Aristotle. In his famous fourfold division of the types of natural explanation (*Physics*, II), he introduced an enigmatic fourth category strongly contrasted by him with the other three, i.e., "material" explanation (or "material cause"). Among the explanatory factors in a complete account of the making of a statue out of bronze will be the bronze itself. Admittedly, the form imposed upon the statue, the agent-sculptor, the end in view, will all contribute more, ordinarily, to the explanation of what went on. Yet in a complete account, the bronze cannot be omitted. But now comes a paradox. The bronze, though a "material" for this particular change, yet has itself a form, and in the description of the change by which it itself came to be from another element, the bronze would be form rather than matter. Thus, the "matter-factor" in an explanation of a change will be relative to the change; it is not an "absolute" component. Bronze will be the "matter" of one change, the form of another. By "matter" here is meant that constituent which is taken for granted as somehow given. In art, it is the raw material. But in nature, it would seem that what will constitute the "matter" will be relative not only to the particular process one is trying to explain, but also to the level of explanation chosen. If one is "explaining" the greying of hair in the adult human, one might take as "matter" the human composite, the living cells of that composite, or the physical constituents of the cells. The choice will depend on the theory of aging one chooses; the "matter" of the change will be the "given" (or postulated) elements which the theory uses: atoms, cells, the whole man, as the case may be.

In the context of contemporary physical theory, because of the shift to the empirical and the hypothetical, this doctrine of "material causality" has to be reinterpreted. In any theory, there are postulated elements (electrons, fields...) whose mode of activity is specified and is not under investigation. The question for the theorist is whether these elements will combine to give the results desired. For a theory is always put forward in the context of some specific set of data, or better perhaps, some specific "level" of data. If the data are combination weights of chemical substances, Dalton's theory of combination will postulate atoms as the "materials"; it will attribute to these only a single characteristic, weight. As far as the theory goes, these atoms can be taken as "ultimates," as containing no further structure than their postulated definition requires. But the simplicity or lack of structure of the "materials" is strictly relative to the theory in which they appear, and thus to the data towards which this theory is directed. A later theory, with new data to handle, may substructure the atoms, may postulate electrons or nucleons, for example. These then become the "materials" of the new theory, themselves now "given," i.e., not subject to further question within the context of this theory.[15]

The crucial point is that the "materials" are always relative to a particular theory, and may never be taken to refer to physical entities that are in reality simple or ultimate. If it could be shown that the theory is itself "ultimate" in the sense of not needing modification in the light of any possible future results, one might be tempted to claim the real referent of the theoretical constructs to be simple and ultimate. But it does not seem possible to make such a claim about any physical theory unless we are also prepared to assert the classical rationalist thesis about the nature of science. The history of recent science does not support this thesis.

§ 7. *The Horizons Recede: Relativity Theory*

But perhaps the greatest blow suffered by the older *a priori's* came at the hands of Einstein. Analysis of the foundations of Newtonian mechanics indicated to him (as it had to Mach and others earlier) that the conceptual structure of the *General Scholium* is not coherent. Not only is there no privileged frame of reference for mechanical phenomena (as Newton had conceded), but there is none for optical phenomena either. The velocity of light is the same, no matter what the motion of source or receiver be. The former is true of such mechanical phenomena as sound waves, but the latter is not true of any mechanical phenomenon, so that light is apparently not governed by the laws of Newtonian mechanics (specifically the law of composition of velocities). From this simple beginning flow some radical results for the theory of measurement (measurements of length and of time are relative to the motion of

[15] There are several different ways in which the complex Aristotelian theory of "matter" as a factor in scientific explanation can be interpreted in the contemporary context. See the Introduction to The Concept of Matter in Greek and Medieval Philosophy, rev. edn., ed. by E. McMullin (Notre Dame: University of Notre Dame Pres, 1965).

the measuring instrument, so that for instance, a rod will seem "contracted" to an observer moving at a different velocity). These results are appreciable only if the velocities involved are very large; for ordinary velocities (like that of a car or even the earth in its orbit), Newtonian laws will still hold. Newtonian mechanics now appears as a "special case," a simple approximation valid only at velocities much less than that of light.

Worse is in store if accelerated motion also be included. Here a much more radical critique of the Newtonian system is required. In his second "Law" Newton had implicitly defined acceleration as an "absolute." But acceleration has to be measured against some reference frame, so an absolute reference frame (the "ether" of the later Newtonians) is implicitly presupposed by the entire conceptual structure of *force-acceleration-mass* introduced in the *Principia*. If this reference frame be rejected, a new and much more complex conceptual structure is required. Einstein presented one such in his General Theory of Relativity. This is a far more speculative effort than his earlier account of uniform motions was; many other attempts have been made to formulate an adequate general account of motion, and the question is currently a highly controversial one. But some things are clear. The commonsense motion of "simultaneity at a distance," of cutting the present moment in an absolute way across the universe, will not do. Estimates of the time of occurrence of distant events are radically dependent upon the state of motion of the estimator; since there is no "absolute estimator" or absolute ether, no generally agreed ordering of events-at-a-distance can be given. The older concept of mass also proves incapable of any exact operational formulation, though it may be retained as a useful approximation.

In our context, the point of all this is twofold. First, classical mechanics is seen to be no more than an approximation for special conditions of motion. Even the Euclidean geometry associated with it no longer provides the simplest mathematical connectivity for the new physics of accelerated motions. Thus the aid and comfort given rationalist theories of scientific evidence by the older mechanics is no longer available. There have been attempts once again to find an *a priori* basis for the newer theories, but these have had little support. It is obvious that Einstein's famous equation is by no means the last word on accelerated motion; even the very choice of tensor expressions in it is dictated almost as much by esthetics as by analysis of physical concepts. Besides, it is now clear that the basic concepts of *time*, *space*, *motion*, *force*, as they occur in current physical theory, are by no means identical with the commonsense versions of these concepts, rooted in direct experience, from which a long time ago they began. Their validation must, therefore, depend primarily upon the validation of the theories in which they occur.

To what extent is a theory still likely to develop? Is it not true in some sense that well-supported theories (like the kinetic theory of gases) are not likely to be replaced except by a wider theory which contains them as a "special" case, so that they would seem to have a definite kind of permanence? This "correspondence principle" has been of great use in recent physics; it specifies useful boundaries for theoretical speculation. Had relativity theory not included Newtonian mechanics as a special case for small values of certain parameters, the new theory would obviously have been unacceptable. There is a complicated question here about the sense in which one theory can "contain" another, and what correspondence really comes to. Certainly the model used by an earlier theory is often dropped by the later one. But it does seem possible to trace a progression in the history of the theory of a given domain (e.g., optics); there is a simultaneous generalization and sharpening in which simpler structures are incorporated in later, more complex, ones.[16]

Modification of a theory can come about for very many

[16] See Mary Hesse, Models and Analogies in Science, Rev. edn. (Notre Dame: University of Notre Dame Press, 1966); also the last section in my "Realism in Modern Cosmology," Proc. Am. Cath. Phil. Assoc., 29 (1955), 137-150.

reasons. The commonest ones are: 1) the extension of the experimental context, e.g., more accurate data, a different range of the parameters; 2) the discovery of a new resonance between two domains of enquiry; 3) the critique, on internal grounds (coherence, mathematical presentation), of the theory. Keeping these in mind, if one asks whether General Relativity Theory is the last or next-to-last word, the answer is: certainly not. It is true that we do not expect any new data in certain ranges; the inverse-square law will certainly not be modified for our planetary system. But if we think of theory (hypothetical, explanatory) rather than law (taken as the statement of an empirical regularity) it is clear that the general theory of accelerated motions has a long road ahead. Much relevant data for the theory is yet to come from such sources as radio astronomy; these will not give us new planetary laws but they will provide a new explanatory context for gravitational motion as a whole. Also there are other relevant domains of theory that have to be brought into coordination—so far, for example, a satisfactory relativistic quantum theory is not available. When one is found, it is certain that quantum theory and relativity theory will each have been modified by the other, and that the "pure" theories will then only be used as approximations for special cases. Lastly, the development of new mathematical formalisms (like group theory) allow reformulation and improvement of theories that are already doing an adequate job.

To put forward a particular physical theory as definitive, then, is a hazardous thing. If it be applicable only to a very limited area, there is always the likelihood of an unexpected resonance developing with some other area. If it be very general, an increase of generalization due to an extension of the data is almost always possible. One might say, for instance, that the electron-orbital theory of the atom is definitive for the entire range of spectroscopic data known. But what about new possibilities of testing parameters in ranges not yet tried? And what about the likelihood of the theory itself being reconstituted in later forms of quantum electrodynamics? The basic concepts of the older theory would then have to be understood in a somewhat different way. Thus, there are many barriers in the way of formulating a definitive equation of motion which will cover all contexts. To the extent that it is empirical, it would seem there is always something more left. Only if it can be insulated from the contingencies of future experiment and conceptual development, can it be asserted as definitive. And this would seem to lead us back once more to the old rationalist hope of a world-formula that would commend itself mainly by its own intrinsic coherence and plausibility. The reasonableness of such a hope is critically dependent upon one's metaphysics and natural theology. Insofar as one looks at the history of modern science as a guide, the issue is still an open one, although for every one Eddington, there are hunderds of more empirically-oriented scientists.

§ 8. *The Horizons Recede: Quantum Theory*

Relativity theory is still in most respects a "classical" physical theory; it is a theoretical extension of Newtonian mechanics into new ranges. But with quantum theory, we encounter a physics so radically novel in inspiration that it deserves to be marked off sharply from what went before. The first quantum theory (1901) was concerned only with radiation, and though its conceptual implications were far-reaching, this was only fully appreciated when a more general quantum theory of matter (1924) was advanced. The major consequence of the quantum theories for our theme was the realization that basic physical metaphors (*wave*, *particle*, *motion*) can be extrapolated only with great caution in the realm of the very small, where the guidelines drawn from middle-sized perception no longer may hold. There is a remoteness, a lengthening of the line of inference, that makes theory itself a less definitive affair than it is at the middle-sized level.

Of the two prevailing interpretations of quantum mechanics, "indeterminist" (Copenhagen, orthodox) and "determinist" (Paris–Moscow, minority view), it is notable that the former is much the more dogmatic about the future of quantum theory. According to one popular version of an argument originally due to John Von Neumann, every future theory of the subatomic domain will not only have to contain the present theory as a limit-case, but also will have to retain the quantum uncertainty principle in its present form. It seems very likely that this argument is circular, and that there is no real evidence for the "orthodox" claim that the present quantum uncertainty is in a very basic sense definitive. The "determinist" on the other hand usually reverts to the Pascal metaphor and sees worlds within worlds in unending (and discoverable) series downwards, "indeterminist" layers supported by underlying "determinist" ones. The scientists who most stress the "infinite future" of science today are probably those physicists (like Vigier) who like to see physics as a progressive unlocking of one world within another.

As a corollary, it might be worth noting that quantum theory has been made possible only by an immense development in the technology of instruments. This technology grows steadily more complex and more powerful. Technology appears to many to have within itself almost unlimited potentialities for development: larger accelerators, more powerful telescopes... The tendency is thus to transfer this unbounded character to the theories which depend most intimately on technology: Theories can no more be definitive than accelerators can. The argument is not always a valid one, but it is certainly true that the rejection of boundaries in the domain of technological possibility influences one to reject any suggestion that a given level of theory might prove definitive.

There can be little doubt that the scientists of today as a group have a better feel for the logical structures of measurement, proof and test, than they had half a century ago. They are more explicitly aware of the hypothetical character of their theories; they are less literal in their devotion to models; they have a better grasp of the hypothetico-deductive character of much of the verification done in science; they have a more exact appreciation of such things as experimental error, statistical analysis, inter-relation of domains, and the rest. They have not learnt this from books about philosophy of science, as a rule. The development has come rather from a professionalization, a more exacting training, within science itself. But though philosophy of science may work primarily as a second-level investigation of what the scientists are already doing, the enormous output of formal philosophy of science in recent decades has surely had an indirect impact on scientific methods too.

This improved "feel" is one last factor worth noting in any account of the latter-day receding of scientific horizons. Science is far more tentative, more modest, less willing to foreclose on its future than it was around the turn of the century. The more empirical a science gets, the less likely it is to seem exhaustible.

PART THREE: MIGHT SCIENTIFIC ENQUIRY STILL BE BOUNDED?

§ 9. *A New Kind of Barrier*

Within this chorus of optimism, it may seem perverse to raise doubts, to ask whether the horizons are quite as far off as people assume them to be. In this final section of the paper, some of the most nagging of these doubts will be briefly sketched. But before this, it is essential to note the similarity and the dissimilarity between the old and the new "limits of scientific enquiry."

In the "classical" period, science tended very often to be regarded as definitive statement. It was assumed to exhaust (or nearly exhaust) some section of the real, and its support came primarily from a penetrating insight into some simple general condition, and only secondarily from an intensive survey of all the variations of the empirically observable that are likely to be relevant to the case. This implies (as we have said before) a definite metaphysical orientation towards the assumption that reality can be grasped in single formulae of human origin. If science terminates in such a view, it would be because there is nothing more to do. Scientific enquiry in some particular domain would end because everything would have been said there. The breakdown of this assumption is parallel with the increase in importance of the empirical; one is much less likely to regard a claim as definitive, if limited experiences have played a role in formulating it. So many are the veils of language and hypothesis that lie between us and the real that we now tend to believe that something exhaustible must be something man-made, or, correlatively, that the real can never be conceptually exhausted.

If possible barriers to scientific enquiry loom on the horizon today, it would not be because science is finished but rather because it could not (in some specific instance) go any further. It is not definitive; it is just as tentative as ever. Even statements about barriers will be tentative, hypothetical. What has happened is simply that the feeling of unlimited horizons characteristic of the early part of our century is being succeeded by a more hardheaded look at the actual prospects. What factors will govern the future of scientific enquiry To what extent will it be progressive as in the past ? Let us take a look first at some results of contemporary theory, and then at the conditions of scientific enquiry itself.

§10. *The Gödel Theorem*

Before we get to natural science, let us glance at one stunning "setback" in mathematics, the well-known Gödel theorem (1931):

> "To those who were able to read Gödel's paper with understanding, its conclusions came as an astounding and melancholy revelation. For the central theorems which it demonstrated challenged deeply-rooted preconceptions concerning mathematical method, and put an end to one great hope that motivated decades of research on the foundations of mathematics. Gödel showed that the axiomatic method...possesses certain inherent limitations when it is applied to sufficiently complex systems...even to the familiar arithmetic of cardinal numbers. He also proved, in effect, that it is impossible to demonstrate the internal con-

> sistency (non-contradictoriness) of such systems, except by employing principles of inference which are at least as powerful (and whose own internal consistency is therefore as much open to question) as are the logical principles... within the systems themselves." [17]

Ever since Euclid's day, the axiomatic method has been held up as the model for all proof. Attempts have been made to axiomatize physical theories like Newtonian mechanics. The main advantage of such an axiomatization would be to make it easier to prove the consistency of the theory—or so it was thought. But now Gödel has shown, by means of a cogent and apparently definitive formal proof, that the consistency of any formal system as complex as, or more complex than, arithmetic cannot be proved. Even more unexpected was the discovery that the completeness of such a system cannot be proved either, which means that no matter how we axiomatize a formal system, there will always be "true" statements in the domain of that system which cannot be derived from the axiom-set chosen. There is thus a radical barrier to complete axiomatization; the equating of mathematical truth with deducibility from a specifiable axiom-set is no longer possible.

Here, then, is a unique limitation, and in that domain—mathematics—where limitations might least be expected. For in mathematics we might seem to be dealing with a wholly exhaustible construction, something that could be fully explored and specified by us since it is our construction. Gödel shows that this is not so; the domain of mathematical truth is just as inexhaustible as the domain of physical truth seems to be. It is not the domain that is limited, then, but our means of organizing it deductively, or more basically, our ability to prove that the constructions we make are consistent. During the past century, inconsistencies have been detected in various parts of logical

[17] Ernest Nagel and James R. Newman, "Gödel's Proof," in: The World of Mathematics, ed. by J. R. Newman (New York: Simon and Schuster, Inc., 1956), III, 1669.

and mathematical practice. It is essential to eliminate these, yet it now appears that it is not possible ever to be sure we have done that successfully.

Such a conclusion was utterly unforeseen. The whole progress of mathematics in the past century or so could be interpreted as a sort of liberation from preset limits: Instead of a unique single geometry whose axioms were true in some absolute sense, alternative non-Euclidean geometries gradually came to seem acceptable as "mathematics." No one asked any more about the truth of the axioms of a proposed system: since mathematics was now understood to be a free creation of the human mind, one could only require that it be consistent. This "formalist" approach, fully stated by Hilbert, forced mathematicians to rely only upon relations explicitly given in the axioms and rules of their system. Though these axioms might make use of words like "line" derived from ordinary usage, no recourse could be had to the surplus meaning of such terms. Intuitions based on extrasystemic knowledge could no longer be relied upon. In the past they had guided discovery, but now Hilbert argued that such intuitions, because of their physical origin, only restricted the mind, and tended to make it overcautious when new and paradoxical generalizations (like negative numbers or non-Euclidean geometries) are proposed.

In Hilbert's new vision of mathematics, the only touchstone of the mathematician's work would be its consistency; there would be no external criterion of any sort. In classical mathematics, on the other hand, the consistency of what was done was implicitly guaranteed by the original extrasystematic reference of such concepts as line or number. There was a physical model for Euclidean geometry, the world of experience, and it could not but be consistent. But now that such criteria are discarded, and the intuition is set free, as it were, the danger of antinomy is far greater. For the "new" mathematics, Gödel's theorem spelt nothing less than a disaster. The single criterion retained by the "new" mathematician is shown to be forever incapable of decisive application in any system of interest. This does not mean that mathematical research is blocked, or even made more difficult. Rather it shows that in this conception of "mathematics," certain highly desirable goals are permanently out of reach.

§11. *Intrinsic Limits in Physics*

As physical knowledge becomes more exact, a variety of limits begins to emerge. There is the velocity of light, now thought to be an upper bound on any physical motion. The Special Theory of Relativity strongly suggests that no form of energy can be transmitted faster than light, and that light itself has a maximum velocity. If this is so, many consequences follow. The most recently discovered far-off galaxies are moving away from us with speeds that get nearer to that of light the further away they are. Are there reaches of the universe that our thought can never compass? We have always known that we cannot reach out to the past, though it reaches out to us. It now seems that even our present is more limited than we had realized; there are unimaginable abysses that we cannot reach and that cannot send any message to us. The science fiction writers cheat when they put their spaceships in "overdrive" or use their "space-warp" to overcome the harsh reality that makes the nearest star four years of travel away, even if we are to travel at the maximum speed that any physical reality can, i.e., that of light.

Looking in the other direction, towards the very small, another sort of problem is emerging. The experimental probing of the nucleus takes vast quantities of concentrated energy. As we go downwards in scale, stability is greater, and it becomes more and more difficult to separate constituents. The macroscopic effects of atomic structure (color, melting point, etc.) can all (or nearly all) be explained without going further in the structuring of the atom than the nucleus. The evidence that allows us to go further than this and ask about electron or nucleon structure is itself of a highly sophisticated kind; it is produced by our machines, and only occasionally by such "natural" events as cosmic rays. As time goes on and machines get bigger and more complex, questions arise about the focusing of energy on tiny areas, and it is obvious that such focusing cannot be indefinitely improved. There is a limit to the amount of energy available, and to the means by which we can bring it to bear. It is not a sharp limit, of course, but it does remind us that the picture of physicists probing ever deeper into the worlds within worlds runs into problems at the very first step: getting the data.

A better-known limit is that furnished by the quantum uncertainty principle. This principle has many formulations, but they reduce to two. One relies upon an analysis of measurement and the interaction it necessarily causes. Because of the "granular" structure of this interaction (symbolized by Planck's constant, h), it can never be exactly calculated, and so every measurement, at any level, is accompanied by a disturbance to previously known parameters of the system, one whose amount cannot be exactly calculated. The other form of the principle is more fundamental, and is rooted in the quantum formalism governing the state-description of subatomic entities. Because the parameters are mathematically noncommuting, certain pairs of them cannot be sharply defined or predicted. If one is dealing with an ensemble, this causes no problem because perfectly definite statistical descriptions and predictions can be given. But if one is trying to handle an individual case, quite basic uncertainties arise. For instance, the quantum state-description of a radium atom could never, even in principle, tell us exactly when that atom would disintegrate. It could only give a probability distribution, at best.

At first sight, it might seem as though this was simply a weakness of the present formalism; after all, it has always been true that a given theory could not handle all the "fine-structure" problems presented to it. When quantum theory is incorporated (as all theories ultimately are, it would seem) in a wider formalism, why should the new theory not be able to give a more exact account of such things as the decay of individual radium atoms? In the forty years since Heisenberg, Bohr, and Dirac first proposed their uncertainty principle, this question has given rise to continuing controversy, and there is still no agreement as to how it should be answered. The "Copenhagen" group from the beginning took their discovery to have a sort of ontological significance. The "uncertainty" in the theory mirrored a real "indeterminism" in nature. It is quite striking to note that this interpretation, which ran so violently counter to the whole spirit of Newtonian science, caught on with physicsts almost immediately, and in a short time became the self-styled "orthodox interpretation." This is all the more surprising in that no compelling evidence, especially none of a conventional scientific sort, could be given for it. It was either supported by extrascientific philosophic views (as in the case of Heisenberg and Destouches), or more often rested on a sort of "hunch." Von Neumann attempted to show in 1932 that all future theories of the subatomic will have to contain an analogue of the uncertainty principle, if they are to account for the data accounted for in the present theory. But it seems agreed today that his proof contains a serious *petitio principii.*[18]

Opponents of the "orthodox interpretation" can be divided into two classes. There are those, first, who defend the opposite ontological thesis, determinism, and claim that further research must reveal the hidden parameters that govern action at all levels, including the subatomic. According to Einstein, Planck, de Broglie, a theory that incorporates an "uncertainty principle" is an imperfect theory by definition, one that confesses our present ignorance. In their view, the predictive uncertainty of quantum theory is no more than a defect of the present formalism, one that all analogies from the past assure us will sooner or later be overcome. Some more recent supporters of this view (Bohm and Vigier) would say, more cautiously, that underlying every "indeterminism" of theory is a "determinism" at a deeper

[18]See, for example, the criticisms of Von Neumann's proof given by D. Bohm in his Causality and Chance in Modern Physics (New York: Van Nostrand Co., Inc., 1957), pp. 95-96.

level, and that the future progress of quantum theory will be a continuing dialectic between deterministic and indeterministic theories. Marxist physicists on the whole appear to lean to this interpretation rather than to the "orthodox" one still favored in the West.

But there is another possibility. One can oppose the "Copenhagen interpretation" without necessarily landing in the Bohm–Vigier camp. For on the balance of the present evidence, it would seem that neither side can enforce his view. Either interpretation is open as far as the scientific evidence goes. If one is to be preferred, it must be on philosophical or other grounds. Such grounds can, of course, constitute a perfectly valid warrant for interpreting the implications of a scientific theory, but it is not at all clear that either side in this controversy is able to provide a philosophy of nature that is sufficiently elaborated to carry the weight being put upon it.

Without taking sides in the Bohr–Bohm controversy, then, it is still possible to note one very important consequence for our theme. Classical physics assumed complete predictability in principle; in practice, it was not always attainable, but it was always believed to be available, given enough time. This is no longer the case. We simply do not know whether the physics of the future will restore complete predictability, even at the level of such relatively large-scale events as atomic disintegrations. It may do so or it may not. We do not yet have the "Gödel theorem" that Von Neumann's proof for a long time was believed to be, i.e., a proof of the impossibility of finding a theory that would restore complete predictability. But leaving aside the fact that the majority of contemporary Western physicists appear to believe (without any proof, strictly speaking) that such a theory never will be found, it is most important to recognize that they may be right i.e., the present quantum uncertainty may very well turn out to be a permanent barrier to complete predictability, a feature of all future theories. We cannot be sure, but there are indications that it is quite possible.

Once again, if this be the case, it is not that present theory has exhausted the physical reality, but rather that it has said all that theory is able to say, even though there is much more that we would want it to say.

§ 12. *The Conditions of Scientific Enquiry*

Science is a human activity. What touches man will thus touch science too. The future of science depends on the future of man: each alike could be wiped out in an instant of decision. Apart from this now never-absent possibility, might there be inherent limits to science arising from the fact that it is man's work? After all, the human mind is a finite thing; it is presumably not capable of indefinite development. For science to have an unlimited future, would the human mind itself not have to develop? Progress in science is not merely in extension, but in depth and difficulty also. The lifetime of science is counted in centuries, and most of it is crammed into the last few. At the present rate of growth—the familiar and probably inaccurate claim that 90% of all professional scientists who have ever lived are alive today comes to mind—can we meaningfully speak of science in the year 3000 AD? or 30,000 AD? or 300,000 AD? Man himself has been around longer than 300,000 years, but the imagination cannot even carry us to 3000 AD. Is there any reason at all to suppose—as is so often supposed in recent years—that the growth of science can continue along the exploding curve of the last fifty years indefinitely, provided no universal human calamity intervenes? If the answer is none, what are the most likely limiting factors on the side of man?[19]

[19]Though many books (under such titles as The Boundaries of Science, or The Limitations of Science) and a myriad articles have been written on the topic in recent years, the only piece I have read that seems worth recommending to the busy reader is a very short one by one of the great physicists of our time, Eugene Wigner: "The Limits of Science" in Readings in the Philosophy of Science, ed. by H. Feigl and M. Brodbeck (New York: Appleton-Century-Crofts, 1953), pp. 757-765.

What does scientific enquiry take? It takes memory, motivation, imagination, among other things. Each of them suggests future—indeed, present—headaches. The process of human learning is conditioned by biological and psychological factors. We are only beginning to appreciate some of these, and spectacular progress, not only in learning theory, but also in its practical applications, seems right around the corner. But with all of this, the pace of learning is still finite, and so is the human life-span. The creative years of a scientist are short, but already the paths to the frontier are dangerously lengthening. It is possible to plan educational short cuts, to drop whole areas and get quickly to areas of contemporary concern. But because scientific knowledge develops mainly by adding new "vertical" layers rather than by making "horizontal" forays here and there, this kind of condensation is possible only up to a point. One can omit classical theories of elasticity, for example, in teaching a student quantum electrodynamics, but one could not omit classical dynamics entirely, or else the student will not really understand. To understand complex theories like general relativity theory, a great deal of prior work on vectors and tensors, on dynamical explanation, is indispensable. In a certain sense, the mind has to recapitulate the history of the theory, up to a point at least, in coming to understand it fully. This is especially true if he is not just to understand it, but to play a role in developing it further. The creative mind has to break the rules, make unexpected connections, but to do this a thorough grasp of the rules (and a feel for how they have been "broken" in the past) is required.

It is already the case in domains like physics that experts are completely out of sight of one another, each busy in his own excavation, the sounds of digging in other excavations quite muffled by distance. They will not know each others' theories, though they still can as a rule understand them if they make a great effort. But life is short and effort is precious, and the *Physical Review* keeps doubling in size. Each year hundreds of thousands of papers in physics are published; several thousand of them are likely to be directly relevant to any given area of concern. It is not enough to gesture optimistically to the computer. Information retrieval of physical facts is very helpful, but there is no way of mechanically "compiling" different theories. One has to read them and understand them, and that takes time and energy. And the good researcher needs to know all, or at least most of the theories that could conceivably be relevant to his quest, as well as to have a good sense of their strengths and weaknesses.

It is a problem, then, of both depth and extent. To understand a theory often requires a long learning process; a worthwhile scientific theory can never be immediately grasped by the untrained mind. Science keeps proliferating in all directions; no one can keep track of all the important contributions, even to an area as well-defined as physical chemistry. Yet for effective work in the area, one has to be able to relate; there has to be a clash of unexpected elements. We stand aghast at the difficulty of reaching the frontiers today; very few educated people do (or could, without considerable further training) understand what is going on today in quantum field theory, for example. But here we are, after only a few decades of really intensive exploitation of science, appalled already at the problems of understanding, or even finding out about, what is going on—and yet we speak glibly in terms of centuries ahead! There is, it might well seem, something facile about such an optimism.

That scientific enquiry depends critically upon the motivation of its practitioners hardly needs to be said. It is abstract, demanding, lonely work with little of the immediacy and warmth of other human activities. It makes enormous emotional demands; it sometimes calls for unusual sacrifices of the most "human" things in life, evenings with the family, leisure with a good book... Of course, it can be done as a "job," and as scientific personnel in government and industry multiply, this will probably become more and more common. But the creative scientist, the one who is likely to push his science into new reaches, is separated from the life of his fellow citizens by years of decisions that others, as gifted as he, have been unable to make. Out of the millions who graduate from high school in the United States, a mere couple of hundred reach a Ph.D. in mathematics each year. And as everyone knows, even a Ph.D. is no necessary proof of creativity.

When one thinks of the vast structure of industry and education in the United States, the crucial role played in both by mathematics, and the desperate shortage of trained mathematicians despite the financial inducements that government and graduate school have been offering, it leads one to conclude that creativity does not necessarily respond to social need. There are undoubtedly thousands out of America's millions, who have the intellectual ability to become first-rate mathematicians. One of the reasons why more do not is surely the strong, almost overpowering, motivation the student must have in order to take upon himself this kind of isolated absolute effort, so unlike anything his more easygoing neighbor has to encounter. The demand for such effort comes especially in late adolescence and early manhood, and it is just at that period that the student of today finds himself increasingly under emotional pressure from a society which in so many ways challenges and dissipates intellectual effort.

Some of the motivation impelling students who take on the years of preparation necessary for advanced work in today's science surely comes from the strong feeling that science can make over the world and relieve its wants. As these wants come more and more to be fulfilled, and science goes off along less immediately serviceable ways (to the moon, for example), will the motivation hold up? The good physics student feels from the beginning of his work something of the excitement that has urged men onwards to know ever since the first astronomers left their warm beds for lonely hours of bleary-eyed measurement. But as physics becomes more abstruse and the moving frontier further off, will this continue to be so? It is not at all clear that it will, and creativity in such a case may not answer the command of government to carry the curve of scientific research upward to new and more distant heights.

§ 13. *Analogy and Discovery*

Scientific progress is dependent most of all upon the imagination of individual scientists. In conclusion, then, it may be well to take a hard look at the conditions of scientific discovery, and to ask whether any of them suggests a built-in limitation. Discovery in physical science has not been just of one kind, one can discover all sorts of things: facts, hypotheses, regularities, concepts; the notion of discovery itself is rather different in each case.[20] To discover a regularity in a set of data, it might be enough to run the data through a computer, though there will of course be elements of personal decision in the choice of data and in the definition of curve criteria to be programed into the computer in advance. Even in the discovery of empirical regularities, then, rule-bound mathematical formalisms play the role of necessary condition, never sufficient condition. Nonmathematical unformalized skills and insights, oriented specifically to the physical order and trained by years of familiarity with that order, will be needed in the discovery of even the simplest physical law.[21]

This is even more true of discovery in the realm of hypothesis and concept. Newton's great achievement was not just to discover novel numerical correlations, but to construct a network of complex physical concepts (*mass, acceleration, force*. . .). The syntax of this network was provided by the differential calculus, but its reference to the physical order, that which made it physics rather than

[20] More fully in "Freedom, Creativity and Scientific Discovery," pp. 125-126.

[21] This theme has been emphasized by Michael Polanyi in his Personal Knowledge (Chicago: University Press, 1958), and especially in some more recent articles such as "The Logic of Tacit Inference," Philosophy (Fall issue, 1965). See also N. R. Hanson's Patterns of Discovery (Cambridge: Cambridge University Press, 1958).

mathematics, was of a far more complex kind. The concepts he used had already a physical meaning of sorts given by ordinary usage. Guided by these meanings, he provided an operational linkage between the conceptual system as a whole and the data of measurement. The analogies leading him from the traditional philosophical concept of matter to his own quasi-operational concept of mass were of a specifically physical sort.[22]

Since this is a crucial point, let us press it a little more closely. A purely formal system is a dead system; it has no resources for further development, since the symbols are defined exclusively in terms of their internal relations with one another. There is no dynamism, nothing that will force a change. Only some sort of "surplus value" in the meanings of the terms can guide significant development. Where the system is a mathematical one, the terms have a "surplus value" deriving originally from the experience of multiplicity, space, and the like, but now developed to a far more sophisticated level. The notion of group, for example, is not exhausted by some formal definition: the notion has shown itself capable of extension and generalization, following the characteristic lines of mathematical analogy, just as the concept of number did at an earlier stage. Even in the most formalist of logical systems, one will usually find a handhold for future extension in the broader categories of inference that the system is supposed to explicate. Lacking such a handhold, discovery is reduced to external manipulation (as in finding the shortest single axiom for a given system, for example).

In a physical theory, likewise, some terms must have a surplus value, a resonance that has not as yet been made fully explicit, or else the system is inert, unable to meet fresh challenge. It was from the surplus value of such concepts as *space*, *time*, *force*, that Einstein reconstructed mechanics in a new style. These concepts left room to move;

[22] Discussed in my paper, "From Matter to Mass," Boston Studies in the Philosophy of Science ed. by R. S. Cohen and M. W. Wagtofsky, New York: Humanities Press, 1965, pp. 25-45.

not only that, but they themselves implicitly suggested the direction the movement might take, so to speak. From the internal explicit formal structure of Newtonian mechanics alone, the new theory could not have been derived, not even with the aid of new results of measurement such as were given by the Michelson–Morley experiment. Not all scientific discoveries involve basic conceptual shifts, but the ones that do are the crucial ones for the scientific enterprise as a whole. And in them the physical reference is what ultimately provides the "surplus value."

This can be seen even more clearly in the case of models. The physicist uses models all the time in his theory-construction. They play a central role in discovery—just think of Bohr's model of the planetary atom, or the Crick–Watson model of DNA. The "model" in each of these instances is a spatially differentiated structure, in which elements whose properties are postulated are related spatially and dynamically with one another.[23] The model there precedes the theory. Indeed, it is from the model that the theory is derived, even though once derived, the theory takes scientific precedence. The model is not a mere summary of the data, it goes far beyond them. It is this "beyond" that must be emphasized, as well as the intrinsic resources that the model seems to bring to the aid of discovery. When Sommerfeld was trying to explain some second-order effects in the hydrogen spectrum, he modified the original circular orbit proposed by Bohr and made it elliptical, thus accounting for the anomalous results. Had he merely had Bohr's theory as a mathematical formalism, there would have been no reason to try this hypothesis. He saw the electron orbit as an approximation to physical reality, and this led him to regard the electron as an individual entity in a definite orbital path. Another example of a fruitful sort of model is furnished by the notion of spin which has guided so many of the fundamental discoveries in quantum theory.

[23] The word "model" has a good many other senses in physics and mathematics, but this seems to be its basic sense for the physicist. See M. Hesse, *Op. cit.*

If our thesis is correct (i.e., that models and concepts with a root-sense in experience have played a central part in past scientific discovery), then it may well be that the most threatening barrier to the future of scientific enquiry lies right here. As our theories move further into the very large and the very small, their anchorage in our familiar middle-sized experience becomes less and less secure. We are forced to modify them in ways that seem paradoxical: we have to combine the metaphor of "wave" (periodic transmission) with that of "particle" (discrete interaction) in order to understand quantum theory, for example. The structural and dynamic metaphors whose roots of meaning (and thus whose "surplus value") lie in the world of perception gradually thin out as we descend into the world of the nucleus.

Indeed, it is not clear whether the notion of spatial structure can be properly applied to entities like electrons at all. The statistics is not the statistics of individual entities; except perhaps under special circumstances, they must not be regarded as individuals. Yet only individual localizable entities can have structure in the ordinary spatial sense of that term. It is obvious that we are becoming more and more remote from the levels where analogies drawn from the middle-sized world could guide discovery. The basic concepts of classical physical understanding—*space*, *time*, *motion* —are still used in a gingerly way in quantum electrodynamics, but paradoxes have developed, and their "surplus value" is no longer quite so trustworthy.

If one looks at recent fundamental particle theory, one is immediately struck by its almost purely mathematical character. Physics has always used mathematics but this is different. Newton used the differential calculus as a convenient syntax, but the weight of his system lay, as we have seen, in its crucial physical concepts, *mass* and *force*. Though a numerical measure of mass was possible, the concept itself could not be defined in mathematical terms; its "home," the locus of its "surplus value," was in the physical world. But nowadays physicists use group theory not just as syntax, but almost as carrying its own semantics as well.

Calculus did not tell us how planets would move (until we had made some crucial physical assumptions about force), but group theory is almost expected to provide us with a theory of fundamental particles, unaided. It is clear that the "surplus value" at the frontiers of nuclear research today lies much more in mathematics than in physics.

It is also clear that discovery becomes progressively more difficult at this remote level. The traditional source of "surplus value" is drying up, and physicists have to fall back more and more on the reserves of mathematics. They have always had these reserves at their disposal. But they had more, much more, and it was from these nonmathematical reserves that most of the historic discoveries of science have proceeded. Our query is now clear: Can we be assured that the resources of mathematics alone will be sufficient to carry physical enquiry, as properly physical concepts and models become ineffective? To answer this in the affirmative might seem to commit one to a mathematical ontology of a Pythagorean kind (as Heisenberg has already noted). But even if it does not, the question of discovery must be faced: as science penetrates into the nucleus, where will the "surplus value" that we have seen to be so central to past discovery in science come from? Will the human mind, which after all is inescapably bound to the physical order in which it first learns to understand, continue to move freely in the distant realm of the subatomic, freely enough at least to wander onwards as it has in the past?

§ 14. *Conclusion*

Our purpose here has been to raise questions, not to make assertions about what will, or will not, happen to scientific enquiry. It is the custom of science to predict, and the power of one's prediction depends upon the solidity of one's theory. It is surely not unscientific to hazard some predictions about the future of science itself, though the basis for these predictions be far from assured. That they should be negative and not just optimistic reflects not a desire to limit the ambitions of science, but rather an honest attempt to see where it may go.

It is worth recalling once again, in conclusion, the difference between the sorts of bounds and barriers we are here evoking, and the kind that classical science (Greek and modern up to about 1900) thought of. Classical science was rationalistic in its implicit claim, for it supposed that the physical reality could in principle be completely grasped by the human mind; a particular physical theory could express the final word on an entire domain. Thus physical reality was viewed as exhaustible, after the fashion of a given mathematical system (pre-Gödel). The barriers spoken of in these last sections are of an altogether different sort even though their effect would be the same—the halting of scientific enquiry. These are possible barriers to human enquiry that may lie deep in the constitution of the world. In this view, the world is not exhaustible, quite the contrary. It precisely never can be exhausted; there is more, infinitely more, to be known but we may never know it.

The Socialization of Science

James McCormack

James McCormack *retired as an Air Force Major General and Director of Air Force Research and Development in 1955. Before that, he was for four years Director of Military Applications of the Atomic Energy Commission. During the past ten years, he has been Vice President of the Massachusetts Institute of Technology until his recent appointment as Chairman and Chief Executive Officer of the Communications Satellite Corporation. While at MIT, his outside activities included the presidency of the Institute for Defense Analyses and trusteeships in several other nonprofit national service organizations working in science and technology. He has also served on numerous advisory committees to many agencies of the federal government. Few individuals possess such a breadth of experience in the administration of large-scale technical activity for and in behalf of the federal government. His community service in Boston included the Chairmanship of the Massachusetts Bay Transportation Authority. He has also been a director of the New England Telephone and Telegraph Co., Eastern Airlines, The Federal Reserve Bank of Boston, and other industrial organizations. He was graduated from West Point, Oxford University and MIT.*

The Socialization of Science

WHEN I WAS INVITED to deliver this lecture, Dr. Steinhardt made the task of preparation easier, or at least made it sound easier, by saying that he only wanted me to talk about the things I have been doing for the past twenty years, in and out of government—about how the government supports research and development, and about how this support is translated into the doing of the work. From that beginning, I was somewhat rocked when I later learned that the title of my remarks was "The Socialization of Science." To one who was raised among southern Democrats, some of whom are even suspected on occasion of having Republican tendencies, suggestions of socialism do not come exactly easy. To this I might add that I am not a scientist. An engineer once upon a time, but never a member of the elite. But it gradually dawned that Dr. Steinhardt's terminology has in it a strong element of accuracy. Considering the growth of the government's interest in science and the dependence of the research community on government support, "socialization" is not an entirely unreasonable description of the process.

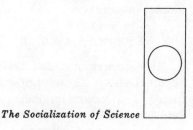

In any case, I accept the title for purposes of discussion. I hope that you will in turn accept a rather broad definition of science, and not be too disturbed if we deal with the interaction of government, science, military and industrial technology, and education across a rather wide range of mutual interests.

THE RESOURCES AVAILABLE TO GOVERNMENT

Our government has traditionally used three means in accomplishing its scientific purposes: government laboratories and arsenals, industry, and the universities. But with the vast changes of the past 25 years, all three of these means need a few words in amplification to couple the past with the present. There are common threads; there are also great differences, even discontinuities.

Historically, the government arsenals performed work which the private sector either could not or would not perform: for example, in chemical warfare and ordinance. Now, large industries have been built for much of this work and are eager for it. Even in the more socially acceptable pre-World War II pursuits of the National Advisory Committee on Aeronautics, the Bureau of Standards, or the Weather Bureau, you will probably agree that the relationship of the present agencies to their predecessors are marked as much by the differences as by the similiarities.

As for industry, some among you may consider that in those distant past days the government simply did not procure research by that route. There is, however, a thread of continuity. When we brought forth a new military airplane in the 1930's, the new art we put into it came from developments made by the airplane and engine companies as well as by the NACA and the military services, and the price we paid the companies carried their continuing effort in technology.

As for the universities, you will surely have been ready to say that twenty-five years ago government support of scientific research was negligible. And it was, but note the exception of agricultural science at some land-grant colleges—an early event in the "socialization" of university-managed endeavor.

In summary, there has been a measure of continuity but the nature of the work and the method of the support have changed, and greatest of all the changes is the size effect of government programs in all three areas. The result is a real difference in kind.

The total of government programs in 1940 cannot be counted up to even 1% of the 15 billion dollars the government spent last year on research and development—or in the current jargon, on research, development, test, and evaluation. Very few, if any, segments of our society escape the effect of these expenditures, one way or another.

THE "IN-HOUSE" ACTIVITY

The government laboratories, as you are perhaps well aware, are today spending more than two billion dollars annually. In the case of both the National Aeronautics and Space Administration and the Department of Defense, the in-house R&D expenditures total some 20% of the R&D budgets of those agencies. Measured as two billion dollars across the government as a whole, or measured as 20% of the innovative effort of a major R & D agency, these are large amounts just in terms of the fiscal attention they require. But they are equally significant in that they support the principal training place of government's expertise for the management of the other 13 billion dollars flowing into the private sector at a crucial point—where the seed corn is planted. In their total effect on the national economy and social development, these dollars are amplified many times, as compared, for example, with dollars invested in repetitive production.

Unhappily, as we all know, the growth of volume of government in-house science has not been matched by increasing quality in the activity. Indeed, all too frequently quality has deteriorated. There are two reasons, mainly. The first is that bureaucracy tends to be incompatible with innovation and the larger the bureaucracy, in or out of government, the less the compatibility in the usual case. The second reason is that the government has been its own severe competitor. By providing to private organizations the means for greater operating flexibility and a more attractive research environment than are enjoyed by its own laboratories, and by generally assigning the more important and interesting work to the private sector, government has left to the directors of its own laboratories an extremely difficult job in maintaining professional morale and attracting new talent.

Of course, it is good for the country and for science that government has provided for a first-rate, over-all national scientific program. But all thoughtful and knowledgeable people in the private sector must necessarily be concerned for the quality of the government's own in-house effort, for its direct productivity in the first instance, but also for its contribution to maintaining the government's expertise for choosing and managing the total national effort supported by government.

We must all applaud the recently intensified efforts in government to improve conditions in the government's laboratories. Three aspects of this effort are worthy of mention. First, professional salaries are being increased, and within reason we will all say: the more the better. Second, more attention is being given to the assignment of interesting and important projects to the government laboratories. And third, modest but not negligible amounts of "discretionary" funds are being given to laboratory directors, to spend where their judgment indicates that the most good will be done. Parenthetically, I have heard that one laboratory director made the first allocation of his discretionary money to improving the janitorial services in the rest rooms. And I have no doubt that he was exactly right in so doing!

Let me offer one additional broad but nevertheless rather concrete suggestion. That is, that there should be a greatly increased effort to find compatible and mutually beneficial ways in which these laboratories can interact with education. I assert as axiomatic that any continuing research operation conducted by or in behalf of government can benefit from association with the rejuvenating, regenerative forces of research conducted in the presence of graduate study. One researcher, one project, one laboratory in one period of time may not need the association, but the continuing operation does. In turn, education at many places can benefit from access to the facilities and programs of the government laboratories, but I would rest the point here on the government's specific interest in its own research operations. We have all seen it work where it has tried with diligence and intelligence—at Oak Ridge, the Bureau of Standards, and the Naval Research Laboratory, to give three examples of modest efforts yielding substantial gains.

So much for the government laboratory. To continue, however, will you forgive me if I remark on a recruiting advertisement appearing in the major newspapers some three weeks ago? It was from the Navy, justly proud of its eight establishments in the stretch from the David Taylor Model Basin to the mouth of the Patuxent River. These establishments along what was called "Research River" employ 4200 professional scientists and engineers and have been a major contributor to the prominent position of the Washington community as fourth among United States metropolitan areas in research and development. Then came the punch line: The attention of the prospective employee was directed to the greater job security these establishments have in comparison with "defense industry."

On this melodious note we turn to the area of:

DEFENSE INDUSTRY

According to the National Science Foundation, the government last year spent 8 billion dollars on RDT&E in industry, exclusive of plant and anything that might still be called production. This figure is important of itself since, as we return again to note, these are the cutting-edge dollars. Being probably a fifth of government expenditures for all purposes in industry, the figure is significant also for what it says about what is happening to the form of industry supporting the technically oriented departments of government.

A study by the Stanford Research Institute of about a year ago gave some interesting statistics on industrial production for government. Apparently, the product of the aircraft industry during World War II sold for an average of 10 dollars a lb. By the time of the Korean War, this figure had risen to 100 dollars a lb. The average product of the aerospace industry today for defense and space objectives sells for an average of 1000 dollars a lb. These numbers say that the scientific and engineering content of the product has risen by a factor of 100 during the course of 25 years. But they say much more: That the structure of this industry has changed in such basic ways as to have a profound effect on its relationships with government and elsewhere in the private sector.

In terms of the visible product of research and development, industry is not now just a production agency, however complex, but covers the entire technological spectrum. Indeed, many industrial organizations are primarily, even exclusively, agencies of research.

Thirty years ago, if government wanted a new cannon there was virtually no choice but to turn to its arsenals, industry having neither the competence nor the interest. As a result, during World War II, when government wanted to develop a major new device such as radar, the proximity fuse, or the atomic bomb, it called mainly on the universities for the underlying development, not on industry. Now industry can not only do these things but jealously guards the entire spectrum as its province. Government's task now is not where it can turn but where it should turn for development work of various categories. Many university research grants, directly a part of the educational process, are paralleled by research contracts in industry, and logically so. Many projects properly assigned to university-managed research centers are paralleled by other projects equally properly assigned to industry. And there is a vast range of research endeavor, indistinguishable by the titles of the projects, which is properly assigned to academic departments, to university-operated research centers, to industry, and to government laboratories themselves.

The government's problem of choice of source is a most sophisticated one. There are few pat answers now and it is unlikely that much better ones will be found except, over time, through better qualified government administration of research and development.

GOVERNMENT ADMINISTRATION OF SCIENCE

We can be more specific, however, about government's administrative methods for dealing with the private agencies. I will comment on three areas: the form of contracting with industry, the translation of new technology into the economic main stream, and patent policy.

As to the contract form, the policy over the years, and the habit of procurement agencies, has been to buy new developments from industry through the medium of contracts which reimburse costs and add a fixed fee, commonly known as the CPFF contract. Granted that the costs of doing new things are difficult to estimate in advance, the CPFF contract has nevertheless too often served to the disadvantage of both government and industry, and acted as a deterrent both to the execution of particular programs and to developing industry's technical competence.

The CPFF system has cost overruns built into it, often because of budget pressures on the representatives of both government and industry to keep original estimates as low as possible. The system has also encouraged the stockpiling of engineers, resulting in artificial shortages of engineering talent. In addition, it has encouraged the assignment of the best of the available talent to proposing new work rather than to performing work previously awarded. The most baleful effect of the CPFF system, however, has been its long-term depressing effects on profit—the life blood of progressive industry. Until this last year, the system differentiated very little between the rewards for excellent and for poor performance and created strong forces to reduce the relatively small but politically vulnerable profit fraction of total costs. Profits in the aerospace industry had reduced by a factor of two over the course of six years: to 2% on sales or 6% on invested capital. The investor could do better in bonds or real estate and he knew it, to the painful disadvantage of the aerospace industry. The more solid of our companies were living on their reserves and the financially weak had suffered worse fates. At best, the industry was inevitably heading toward becoming a ward of the government.

Then, to its lasting credit, the Department of Defense saw the necessity for turning this situation around. Increasing use is now being made of the fixed price contract and of the contract for cost-plus-an-incentive-fee, called CPIF. The former can be a cruel instrument, but it reaches directly to the source of industry's greatest strength—competition in performance. The latter at least increases the reward to the good performer and exacts some penalty from the mediocre. Happily, we seem now to be moving in the direction of restoring the spirit of private enterprise by the traditional means of competition, rewards, and punishments. In so doing, I am convinced that we will reduce total costs.

In the translation of new technology from government programs into the main stream of the national economy, NASA deserves a special accolade. The problems are difficult. For example, the materials, the design philosophy, the end objective, even the general technology which makes the best nuclear reactor for a submarine does not lead obviously to economical commercial power from fixed generating stations. As another example, in space electronics where performance and reliability are worth almost any price, the products for government do not find ready markets in commercial use. Even the organization and methods for doing business with the government can be far apart from those which succeed in consumer product industry.

The process of translation is nevertheless of crucial

importance if the public is to receive its just due from these national programs. NASA has no monopoly in the field, but let me mention as one example of courageous experimentation the agency's 150,000 dollars grant last year to the University of Indiana matched by industry, the purpose being a systematic exploration of NASA's research and development records for potential applications to the general economy. This particular effort may not ring the bell, although we should hope that it does, but it points toward the sort of imagination and ingenuity which we must apply not only to the government programs themselves, but also to capitalizing on them for the total national benefit.

Third, I mentioned government patent policy, where it seems abundantly clear to this observer that despite an occasional exception there is a strong trend toward defeating the general public interest and even the specific interest of government. I refer to the idea that patents emerging from government-financed research and development should generally and automatically belong to the government.

I believe that I am aware of most of the major deficiencies in modern application of the traditional patent system which derives from the Constitution. The need for thorough review of the system is obvious. But the superficially simple and logical solution of government ownership is not the right answer. It would substantially remove private incentive to develop the new ideas into general application. This is more than just antithetical to our free enterprise system, it is directly detrimental to the interest of the taxpayer footing the bill. With government participating in three-quarters of the most important research and development in the country, the stakes in this issue are enormous.

RESEARCH CENTERS

The large research center operated under contract for the government is pretty much a novelty of our era. The prototype was supplied by the World War II laboratories of the Office of Scientific Research and Development. Some early examples were the Metallurgical Laboratory at the University of Chicago, the Los Alamos Laboratory of the University of California, the Radiation Laboratory of MIT, the Applied Physics Laboratory of Johns Hopkins, and the Jet Propulsion Laboratory at the California Institute of Technology. Postwar examples managed by industry are the Knowles Atomic Power Laboratory and the Bettis operation of General Electric and Westinghouse, respectively, for the Atomic Energy Commission. Others are the Oak Ridge National Laboratory operated by Union Carbide, and the Sandia Corporation operated by Western Electric. Another group of organizations falling loosely into the same general category are the special, national service, nonprofit corporations managed by independent boards of trustees. There are some eight or ten such corporations to which we might accord genuine national significance, such as Rand, the Institute for Defense Analyses, Aerospace, and Mitre.

In FY63, the government spent 1.25 billion dollars through these contract research centers: 668 million dollars through those managed by universities, 435 million dollars through those managed by industry, and 125 million dollars through the special "nonprofits."

All of these organizations were established initially to meet an important need of government, which the responsible officials of government saw no other satisfactory means of meeting. On the record, all of them were initially successful in accomplishing their missions—several of them spectacularly so. It is fair also to say that where the initial mission has run out, most of them have transformed into useful continuations in their fields of competence.

It is often asked whether they are now pretty much fixtures, and the answer in most cases is that they probably are. Parkinson's law works here too, although not so strongly as in the case of an organization directly a part of government. One reason for their durability, of course, is that they represent bodies of organized competence the creation of which cost a great deal of effort and money, and the continuation of which, at almost any point in time, is in the government's interest. It must be noted, however, that very few of them rest entirely easy within an area which might be symbolized as a triangle, the apexes of which represent the traditional sources of national strength: government, industry, and the universities. The occupants of the traditional apexes may be expected indefinitely to exhibit some measure of hostility whenever the "interloper" seems to come too close to being a competitor.

From considerable experience with these organizations—and as an almost professional discussant of their anatomy, proper utilization, and future—I seem always to come back to a couple of rather basic truths and another couple of simple but largely unanswered questions.

First, it is entirely clear that the continuing usefulness of these organizations is dependent on their continuing to represent quality. It is not so clear how this essential virtue is to be maintained in perpetuity, considering their relative weakness in providing the stimulations and incentives which preserve the traditional centers of strength in government, industry, and the universities. Continuing, high-level attention on the part of the government clients is of course one key requirement.

Second, it was the government in the first instance—and high officials of government, at that—who called urgently for the establishment of these organizations. In all cases, the call was issued only after thoughtful examination of the government's needs and alternative means of meeting them. In most cases still, the organizations are of great importance for the missions they are now performing. It is quite a lot to expect, however, that the successive generations of high officials of government will continue forever to give urgent attention to the continuing need for these organizations or for their role and welfare.

One of the two questions which seem always to be a residual of such a discussion as this is whether these organizations, in some cases called "National Laboratories," can in fact become really national. Can the organized talent, once it has served the essential purposes of its initial sponsor, be redirected to meet a major, urgent need of another agency of government? Can the habits of government permit, for example, a shift of sponsorship from the Air Force to the Federal Aviation Agency, or from the Atomic Energy Commission to the Weather Bureau? Knowing that I spent many years in bureaucracy, you will no doubt think that my question is cynical, and perhaps it is to some degree, but not altogether.

The second question repeats an earlier point. Is it not possible for these research centers and other special technical organizations, together with their government sponsors and the universities, to find more productive mutual ties with education—for the benefit of all concerned?

ACADEMIC RESEARCH

Turning finally to the universities proper and to the relationships between teaching-oriented research and federal sponsorship, we come to the topic which most of us will agree is the most important of those on our agenda.

The country has come a long way in recent years in clarifying and understanding the issues involved in relationships between the federal government and higher education. Only five years ago, when MIT was approached by the American Assembly of Columbia University to prepare a paper on federal sponsorship of university research, the reason given was that no other educational institution was in quite so good a position to make valid comment on this tangled subject. And such indeed was the reluctant and somewhat unhappy conclusion of the two authors who undertook to produce the paper when we discovered how much "basic research" we ourselves had to do to produce a factual paper. It took us the better part of a year to develop the background of the article which some of you may have seen as a chapter in a book published by the Assembly in 1960 entitled, "The Federal Government and Higher Education." Now I find on my desk in one week two treatises on the subject, each better than the article to which I refer, and each by its multiple authorship clearly indicating how widespread is present knowledge and understanding of the issues. One of these is "New Prospects for Achievement" published by the American Council on Education, edited by Harrison Sasscer. The other is entitled "A Statement on Federal Relations" issued by the American Association of Universities.

As in the case of government's interest in science generally, government's interest in education has a thread of continuity as old as our republic itself. May I quote from two Congresses of long ago. First, from the Northwest Ordnance of 1787: "Religion, morality, and knowledge being necessary to good government and the happiness of mankind, schools, the means of education shall forever be encouraged." That statement of high principle was given practicality be a grant of two townships to the Ohio Company to gurantee a future university in that state. The second quote is from the Morrill Act of 1862, formalizing the system of land-grant colleges. There, the stated purpose was "the liberal and practical education of the industrial classes in the several pursuits and professions of life."

THE GOVERNMENT AND EDUCATION

Yes, the interest of the federal government in higher education has existed for a long time. What caught us somewhat by surprise was the great surge of federal research funds in the universities which began some 20 years ago and proceeded for too long with too little meaningful consideration by either the government or the universities of the total impact on education. In fact, we might say that philosophical attention by the Congress to the broad problem of education began only with the National Defense Education Act of 1958 which initiated support for education in science, mathematics, and foreign languages. Subsequent legislation has made available loans and grants for a wide variety of scholarships, fellowships, and other university purposes, including construction. Even the humanities and social sciences are now receiving some notice.

Federal funds now supply a fifth of United States university budgets. Research of course is still the largest part. The federal government now spends three-quarters of a billion dollars a year on its research programs in universities, exclusive of the operation of university-managed research centers. This amounts to three-quarters of the research bill in the universities.

This is the background for worry that as federal funds loom ever larger in university finances, federal control of education is following closely behind. We agree that the money is necessary, but we ask even so whether it is good. I argue that it has been good, still is, and can continue to be. Our problem is to shift mental gears in pace with the shifting financial gears, to insure that educational policy—and not accounting procedures—continues to control the means by which the federal funds are injected into the educational system.

May I quote from a statement of September 26, 1963, by Dr. Max Tishler, a respected and thoughtful director of industrial research.

> "The Federal funds now flowing in through the windows of university laboratories have created problems that are far more subtle than the ancient one of outside dictation. They include the resulting imbalance between teaching and research, between basic and applied research, and between science and the humanities; the disproportionate growth of the physical sciences; and the rising quality of the best institutions which are those most favored with funds at the expense of the spreading mediocrity of the rest."

I would only change Dr. Tishler's tense and readjust his emphasis slightly. These have indeed been forces deserving of our concern, and they continue to be. But a matter for concern is one thing, and an accomplished disaster is another. The disaster has not occurred and need not occur if we remain alert to avoid it.

For the record, let us note the absolute absence of examples of "outside dictation" despite almost a century of federal support of education in one form or another. Let us note also that the imbalances—which do exist and should be remedied—are in substantial degree no more than a matter of administrative restrictions imposed by the Congress and the Executive Departments on cost reimbursement. In my opinion, these have mainly resulted from the lag between the understanding of the need for federal research support and an understanding of the means by which it is most equitably and most effectively supplied. That this latter understanding has been slow in arriving must be laid in part on the doorstep of the universities, who have been late in analyzing the new body of relationships between education and the federal government induced by the increasing and entirely well-intentioned federal interest, first in research, and more recently in education more broadly. At least the deficiencies in government administration, and in the practices of the universities as well, which are the the direct and largely unnecessary source of much of the present discomfort in the universities, are being set forth with increasing clarity and force—which in our democratic system means that they have a chance of being cured.

One more quote in closing, this from the President's Science Advisory Committee.

> "Whether the quantity and quality of basic research and graduate education in the United States will be adequate or inadequate depends primarily upon the government of the United States. From this responsibility the Federal Government has no escape. Either it will find the policies–and the resources–which permit our universities to flourish and their duties to be adequately discharged–or no one will."

That judgment was published five years ago. Its injunction is now the policy of the Congress as well as of the Executive of the United States. Under our system, and considering the complexity as well as the gravity of the subject, a five-year phase lag is pretty close coupling.

Practical Uses of Atomic Energy

Arthur E. Ruark

***Arthur E. Ruark**, a Johns Hopkins man, with a scientist wife, has been in a variety of activities. He worked on oil production at Mellon Institute and Gulf Research Laboratory, and in over twenty years of teaching and academic research he has been at Yale, Pittsburgh, North Carolina, and Alabama. At North Carolina he held a Kenan Research Professorship and at Alabama was Temerson Research Professor. After World War II he was with the Johns Hopkins University Applied Physics Laboratory, and was then Assistant Director of the Institute for Cooperative Research, taking care of contract matters for the University. He became interested in the search for fusion power during summers spent at Oak Ridge National Laboratory and this led on to his present position, Senior Associate Director, Division of Reseaarch of the U.S. Atomic Energy Commission.*

Practical Uses of Atomic Energy

FOR A MOMENT, let us reflect on the structure of science, in a very broad way. Up to nearly the end of the last century, our science and technology were based on the parts of mathematics and physics, chemistry and biology, which deal with whole atoms or whole molecules, or still larger pieces of matter. When electrons and ions came into a state of respectable acceptance, scientific people realized that they were dealing essentially with electromagnetic forces.

Then, in a few short years, three things of everlasting import happened. I refer to the discovery of X rays by Rőntgen; the discovery of radioactivity by Henri Becquerel; and Planck's great intellectual feat, the revelation of quantum physics. Then the floodgates were open. In a short time it was shown that there were great stores of untapped energy in some elements of high atomic weight. It became clear that a single atom could release an amount of energy a million times greater than the amount released when we combine an atom of carbon and a molecule of oxygen. Evidently, strong forces were at work. As a matter of fact, at least two forces were at work: the strong force between heavy nuclear particles and a very weak force connected with production of electrons and neutrinos.

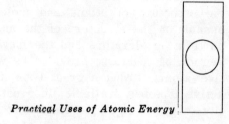

These forces have not been fully elucidated to this day. Nevertheless, the discovery of radioactivity has led to copious fruits. First, it led to experiments which uncovered the structure of atoms and molecules. Then came the assault on the structure of the nucleus. The invention of particle accelerators and the inevitable aftermath, the invention of reactors, provided us with a wealth of radioactive atoms. What a great boon it is to have these radioactive isotopes available, in great variety, in great quantity—eager to sign their names—to reveal their presence in all sorts of experiments. All of a sudden, people could deal with quantities of matter so small as to defy detection by the most sensitive balances or the most delicate chemical tests. All of a sudden, people had the power to see the tracks of individual fast particles, and to detect individual photons.

When some scholar, a million years hence, writes a definitive history of the twentieth century, he may say:

> "That time was a Golden Age of physical science. All that had gone before was employed to study and to utilize the nucleus of the atom. Then the results of nuclear science came to bear on every other field of science and many fields of technology. From this activity arose the proper study of the basic particles of this Universe. Thereby people came to some understanding of the stars, and the galaxies. So that century has become the envy—nay, the despair—of those who came later and looked in vain to reap a harvest half as rich."

Now you will know that I feel a special responsibility, being the only speaker in this series of lectures who was asked to deal with atomic energy. The topic is: The practical uses of atomic energy. But what is the meaning of the word practical? I suggest to you that an action is practical, or a body of knowledge is practical, if it leads to any sort of improvement in our bodily, mental, or spiritual weal. This saying throws the burden on the definition of "improvement" and the definition of "weal." I shall not presume to tell you what they mean. They mean different things to different people but here is my point: The relative emphasis on material needs, and intellectual needs, and spiritual needs, is changing rapidly in your lifetime and mine. What is the use of our material wealth if it does not lead to deep understanding and great peace of mind? Wordsworth said,

> "The world is too much with us. Late or soon,
> Getting and spending, we lay waste our powers."

These necessary things having been said, I invite your attention to the roles played by nuclear science in advancing our total welfare. On the industrial side, I shall speak of the search for uranium and thorium, of the production of nuclear power, of the harnessing of large explosions for peaceful purposes, and the use of isotopes in industry. Then I shall speak of applications in the chemical, biological and medical fields; and finally I shall say a few words on the great problems of nuclear physics, the study of the basic particles and the study of the cosmos. I shall rely quite heavily on simple illustrations, believing that parables will be more pleasing to you than generalities.

In this country, these matters—excepting the astrophysical problems—are dealt with by six divisions of the Atomic Energy Commission: raw materials, reactor development, peaceful nuclear explosives, isotopes development, biology and medicine, and research.[1]

RAW MATERIALS FOR NUCLEAR SCIENCE AND INDUSTRY

The search for uranium and thorium has been very ably prosecuted by the Division of Raw Materials, which also takes care of the activities on milling and concentration of the ores.[2] The intensive search for uranium began with Geiger counters and passed on to the stage where far

[1]The Division of Research takes care of basic research in the physical sciences.

[2]R.D. Nininger, Minerals for Atomic Energy: A Guide to Exploration for Uranium, Thorium and Beryllium (New York: D. Van Nostrand Co., Inc., 1954), 367 pp.

Jesse C. Johnson, "Nuclear Fuel for the World Power Programs," Peaceful Uses of Atomic Energy, 6 (1956), 60. United Nations, N. Y., 1956.

R.L. Faulkner and W.H. McVey, "Fuel Resources and Availability for Civilian Nuclear Power, 1964 — 2000." Proceedings of the 3rd International Conference on Peaceful Uses of Atomic Energy, 12 (1965), 11. United Nations, N. Y., 1965.

more sensitive instruments, the scintillation counters, could be mounted on airplanes for rapid survey of thousands of square miles of rough country. The capacity of the industry increased to meet all current needs, so that now we are going through a period of somewhat reduced production, which no doubt will increase again as the construction of power reactors accelerates in the years ahead.

To provide some notion of the industry which has emerged, I present Fig. 1, a photograph of Anaconda's open mine called Jackpile, and Fig. 2, which shows the Blue Water milling activity. Both are in New Mexico.

POWER REACTORS

Fission Reactors

The first fission reactor came into operation in Chicago in 1942. In a surprisingly short time thereafter, larger

Fig. 1. The Jackpile Uranium Mine of the Anaconda Co., Valencia County, New Mexico.

Fig. 2. Bluewater Uranium Mill of the Anaconda Co., Valencia County, New Mexico.

reactor units were built and successfully operated. Nevertheless, developing reliable and economical reactors for civilian power production is a complex and time-consuming enterprise. The problem of reducing costs has been a main concern through all this period. Allow me to quote a 1962 report[3] of the Commission to the President:

> "Certain classes of power reactors, notably water-cooled converters producing saturated steam are now on the threshold of economic competitiveness with conventional power in large installations in high fossil fuel cost areas of the country. Foreseeable improvements will substantially increase the area of competitiveness."

Figure 3 shows the Dresden boiling water reactor in Illinois, designed for about 180,000 kW of electrical energy.

[3] "Civilian Nuclear Power, – A Report to the President," U.S. Atomic Energy Commission (November 1962). Appendices to that report, bound as a separate monograph, are also available from the U. S. Government Printing Office, Washington, D. C.

Fig. 3. The Dresden Reactor of the Commonwealth Edison Co., Grundy County, Illinois.

This is about enough electrical power for an industrial city of 400,000 persons.

Breeding to Expand the Utilization of Reactor Fuel

The existing power reactors are visible evidence of rapid progress over a period of 20 years, but from almost the beginning one could roughly foresee the shape of events still to come. Looking aside from installations which are definitely experimental, in present power reactors only a small fraction of the nuclear energy available from uranium and thorium is released. Methods for getting out a large fraction of that energy should reach fruition about 1980. Let me describe the rationale of this situation.

Uranium has two principal isotopes, U-235 and U-238. The U-235 is only 0.7% of natural uranium. While fast neutrons can cause either of the two isotopes to undergo fission, only U-235 responds to relatively slow ones. For a number of good technical reasons, relatively slow neutrons are the ones mainly responsible for carrying on the chain reaction in a uranium reactor. In the case of thorium, there is only one isotope, and it does not undergo fission by slow neutrons. Because of these facts, most of the present reactors are fed with uranium enriched with U-235—that is, uranium which has been put through an isotope separation process. In rough general terms, it would appear at first sight that we are operating reactors in such a way that only about 0.7% of the uranium can be burned.

But there is a means of escape from this situation. There are always excess neutrons, over and above the number required to carry on the uranium chain reaction. These can be used—and are being used to some extent—to transmute some of the U-238 into plutonium-239, which can respond to slow neutrons and therefore can be used as reactor fuel. Furthermore, in principle, the neutrons from U-235 can be used to transmute thorium into an isotope of uranium which is conveniently burnable. These two processes are called breeding. If a reactor makes more fuel than it burns it is called a breeder reactor. Otherwise it is called a converter. In principle, if breeding is carried on long enough, a large fraction of the uranium and thorium can be burned. The first breeder reactor was operated in Idaho in 1952.

Figure 4 shows the Fermi reactor, on the shore of Lake Erie between Toledo and Detroit, Michigan. It is the first large-scale breeder reactor. It is confidently expected that continued development will make breeder reactors attractive from the cost standpoint in the 1980's.

Long-Range Fission Fuel Reserves

Estimates of the length of time our fossil fuels will last are quite variable, depending on assumptions as to future population, how much an individual will consume as the years roll by, and other assumptions. I venture the purely personal surmise that use of coal, petroleum, and oil shale as sources of power and heat will become somewhat unpopular in less than 100 years because these accumulations from past ages will be deemed more valuable as starting points for a host of chemical products. Whatever the time scale of these events, the day is certainly coming when nuclear fuel will play a major part in the power economy of this globe. There has been discussion of the possibility that the uranium content of granites could be profitably extracted at some far-off date.[4] To the extent that this comes true, one can say that we have nuclear fuel in quantities which would suffice for the needs of the present world population for many million years.

Feasibility Research on Controlled Fusion Power

The release of energy in a fission reactor is an instance of gaining heat energy at the expense of mass. In the fission process we lose matter, and we gain corresponding energy in the form of heat. Now, energy is also released when many kinds of light nuclei collide with sufficient speed.[5] Such reactions are called fusion reactions. Can this energy release be made to yield useful power? The most interesting case involves the collision of heavy hydrogen or deuterium, with a still heavier form called tritium. In the reaction of these materials, the release of energy per gram of fuel is five times greater than in the case of fission, but it is extremely difficult to create the conditions needed for effective reaction. We have to heat the reactants to a temperature of at least 50 million degrees centigrade before the reaction can become barely self-sustaining. Then we must confine the hot gas until the nuclei react, after a great number of ineffective elastic collisions. This is true because mere bombardment of a target containing tritium, with fast deuterons, leaves us far short of the goal. We have to increase the number of collisions by giving the particles extended opportunity to trade energy among themselves.

[4] A. Weinberg, "Energy as an Ultimate Raw Material," Physics Today, 12, (November 1959), 18.

[5] Amasa S. Bishop, Project Sherwood (New York: Anchor Press [Doubleday & Company, Inc.] 1960), 216 pp.

Fig. 4. The Fermi Fast Breeder Reactor, designed by Atomic Power Development Associates, owned and operated by Power Reactor Development Co.

Such a hot gas, called plasma, cannot be held in check by material walls; only one way is known to confine it. The gas is completely ionized; therefore, the moving charged particles can respond to a magnetic field, which causes them to move on spirals, as shown in Fig. 5. This expedient has enabled us to attain the necessary temperatures—more than twice as high as the temperature at the very center of the sun. We can increase the temperature further, if we need to do so. So one great part of the task has been accomplished. The remaining great problem, increasing the confinement time, is difficult indeed. Hot plasmas are afflicted with instabilities, eddying motions, turbulence. Our laboratories and those abroad are engaged in a struggle to control these unstable motions. No one knows the outcome, though steady progress is being made. In this effort we move with single-

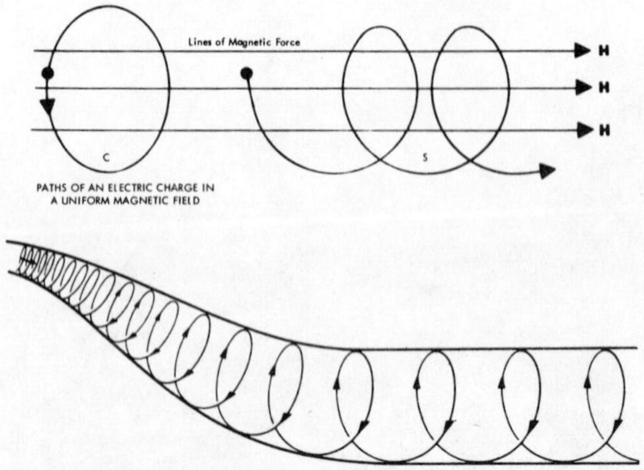

Fig. 5. Motion of electrons and ions in a magnetic field. The field pushes a moving charge at right angles to its direction of motion and also at right angles to the field. If the particle is moving perpendicular to the field, its path is a circle. The field does not affect the portion of the motion parallel to the lines of force, so in general the path will be a spiral like a screw thread. Endwise confinement can be provided by a magnetic field which increases from left to right, as lower figure demonstrates.

hearted purpose to do one or the other of two things: either find the way to controlled fusion power—if that is possible—or else, prove beyond the shadow of a doubt that we cannot do so—if such is the case. Figure 6 shows an apparatus, a magnetic bottle, used for this research, at Oak Ridge National Laboratory.

PEACEFUL NUCLEAR EXPLOSIVES: PROJECT PLOWSHARE

The Commission's work on using atomic explosives for peaceful purposes is called Project Plowshare.[6] There are two categories of these explosions, namely, contained ones and explosions which make an open crater. Contained nuclear explosions have a variety of possible uses. They can shatter rock underground, thus leading to very useful developments both in mining and in oil and gas production. Furthermore, there is the possibility of loosening up the strata to provide water flow where it does not now exist.

[6] G. W. Johnson and Harold Brown. "Non-Military Uses of Military Explosives," Sci. Am., (December 1958), 7 pp.

Neutrons from the explosion are available to transmute uranium, plutonium, etc., resulting in the production of heavier elements, the transuranic ones with atomic numbers higher than that of uranium.

It is important to note that in these explosions the radioactivity is mainly confined to the central chamber gouged out by the explosion and to fissures which do not go very far from that chamber. Figure 7 shows two views of the chamber formed by an explosion carried out in 1961, and Fig. 8 shows how the nature of a cratering event can vary with the depth of burial of the explosives. The idea is to

Fig. 6. A magnetic bottle, called a magnetic mirror. This device, called the Direct Current Experiment Two (or DCX-2) is at Oak Ridge National Laboratory. The massive coils provide a strong magnetic field. The operation is continuous.

Fig. 7. Views of the cavity formed by the underground nuclear explosion called GNOME. (Lawrence Radiation Laboratory, Livermore). The scale can be appreciated by looking for a man near the center in the lower picture.

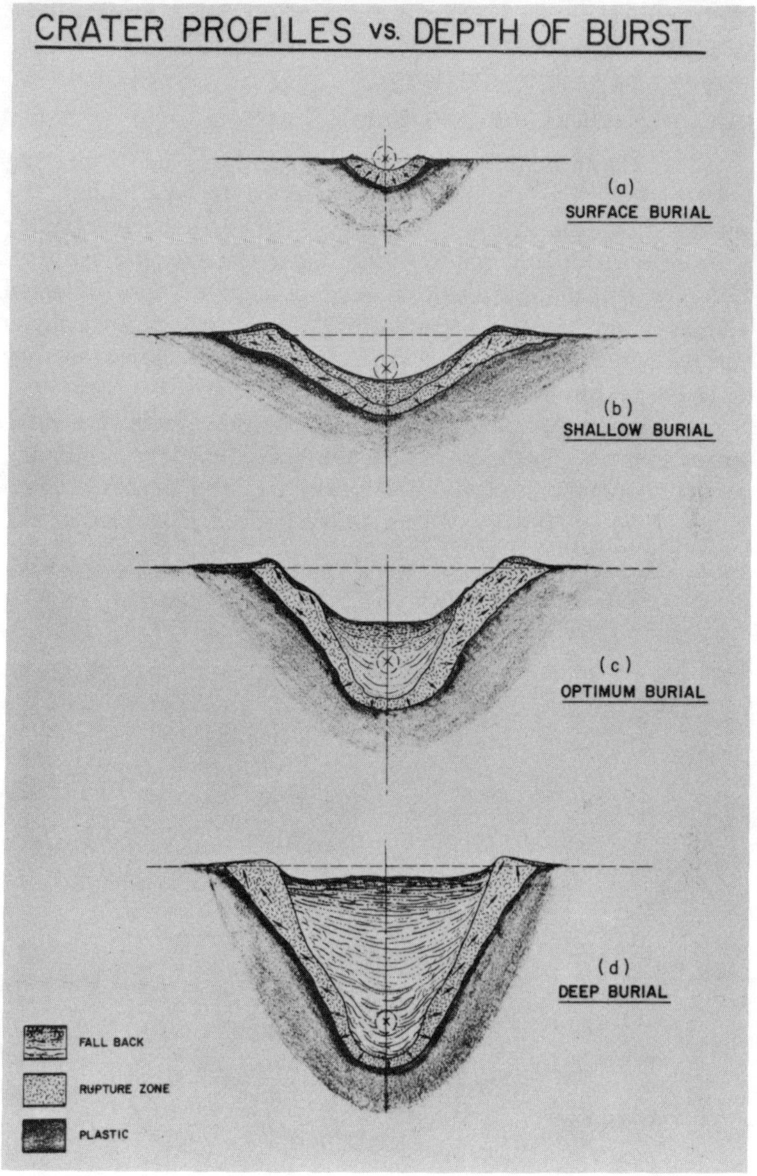

Fig. 8. Cratering profiles versus depth of burst.

dig harbors or to excavate canals, when and if international conditions permit.

ISOTOPES FOR INDUSTRIAL USE

The Commission has played a leading role in making both radioactive and stable isotopes easily available. The marking of materials with stable or radioactive tracer elements is an old game which began early in the century, but now its applications are very numerous. Beneficiation of materials by heavy irradiation came into the picture much later. A few illustrations will suffice to show the versatility of these techniques.[7]

1. Figure 9 deals with tracing the movement of underground water. In the past this has been done by introducing a fluorescent dyestuff. The person in the center holds a flask of water, labeled with a heavy hydrogen isotope called

[7] P.C. Aebersold, Chapter on Atomic Energy Benefits in Atoms at Work (Culver City: Murray and Gee, 1956), 48 pp.

Fig. 9. Use of tritium to replace large quantities of fluorescent dye for tracing underground water flow.

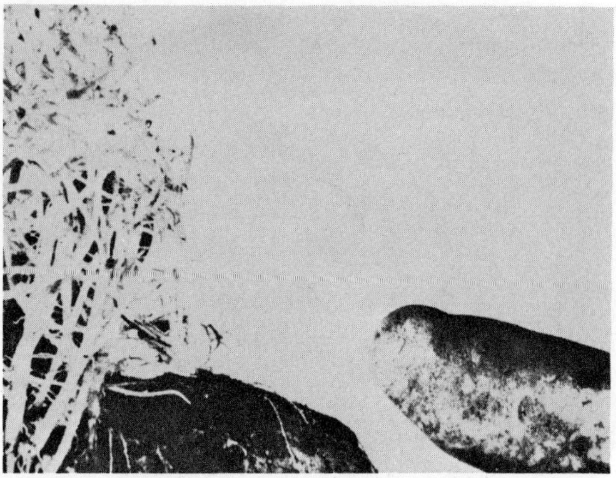

Fig. 10. Irradiation stops the growth of sprouts from stored potatoes.

tritium. That small amount of water can be used to make an investigation which otherwise would require all the organic dye piled up in the background.

2. Another interesting field is the irradiation of food, with suitable sources of gamma rays or other radiations. Figure 10 shows how the irradiation of potatoes stops the growth of sprouts while the potatoes are in storage.

3. Figure 11A shows a very compact X-ray source. The X rays are excited by beta rays (or fast electrons) from a radioactive isotope of promethium. Figure 11B shows an X-ray photograph taken with this apparatus. We can well imagine the value of such equipment to military people or explorers.

APPLICATIONS IN BIOLOGY, AGRICULTURE, AND MEDICINE

A separate lecture would not suffice to give an idea of the utility of tracer methods in the elucidation of chemical and biological mechanisms.[8] The method is a powerful one in such complex problems as the study of photosynthesis

[8] C. L. Comar, Radioisotopes in Biology and Agriculture, (New York: McGraw-Hill Book Company, 1955), 481 pp.

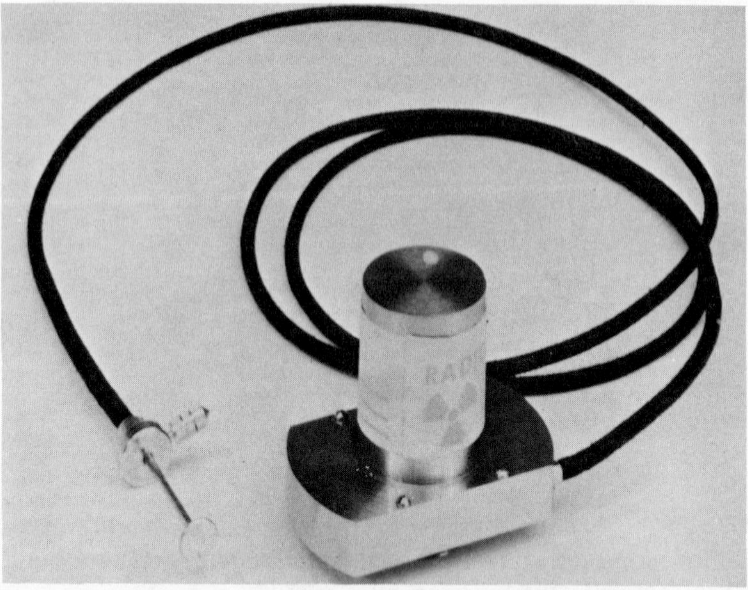

Fig. 11A. Compact beta ray source yielding X rays for field use. Developed by Armour Research Foundation under an AEC contract.

and of metabolic reactions. Passing these matters by in spite of their unlimited implications, let us consider a few agricultural and medical applications.

Genetic Applications

I remind you of studies on mutations carried out by radiation. This method has been a key procedure in genetic research. The use of radiation in plant breeding is illustrated in Fig. 12, which shows the so-called "gamma garden" at Brookhaven. The plants are set in circles and receive various doses from a strong gamma ray source placed in a box in the middle of the garden. By methods like these, we already have peanuts improved by radiation, in the grocery stores. Again, in Fig. 13, we have a demonstration of the use of radiation to produce a rust-resistant wheat. The short sample is wheat afflicted with a rust disease, while the sample at the left with a long green stem is the one improved by radiation.

Medical Uses of Isotopes

It happens that iodine is preferentially absorbed by the thyroid gland and this makes it possible to study hyperthyroidism and other troubles of the gland. In Fig. 14 we have an illustration of the use of iodine-131 at the Oak Ridge Institute of Nuclear Studies to show details of a human liver. The detecting instrument is moved along over the region of the liver and responds to the concentra-

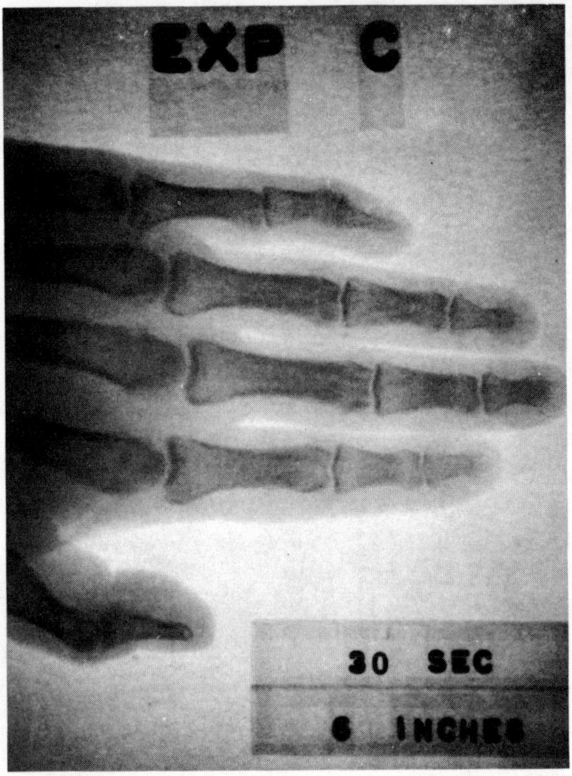

Fig. 11B. A radiograph of a hand obtained with the apparatus shown in Fig. 11A.

Fig. 12. The gamma ray garden for irradiation of plants at Brookhaven National Laboratory.

tion of the radioactive material in particular parts of the organ. Brain tumors can be located in similar fashion. These advances depended on the basic research on scintillation counters which led to successful detection equipment.[9] As to the use of radiation for therapy, I mention only the employment of radioactive phosphorus in studying blood diseases and the very extensive use of radioactive cobalt as a replacement for high voltage X-ray therapy, especially in the case of deep-seated tumors.

BIG ACCELERATORS AND FUNDAMENTAL PARTICLES

Turning now from these mundane things, these material advances, I come to the matter of research on the basic

[9] J. E. Francis, P. R. Bell, and C. C. Harris, "Medical Scintillation Spectrometry," Nucleonics, 13, (No. 11, 1955), 82. Material by the same authors may also be found in Report TID-7673 entitled Progress in Medical Radioisotope Scanning, 1963, obtainable from the Office of Technical Services, Dept. of Commerce.

particles. The last decade has seen great advances in the design and construction of high voltage accelerators, giving particles with energies from 1 to 30 billion electron volts. They enable us to study the detailed behavior of the permanent particles and to create a host of ephemeral ones by collisions of protons, neutrons, electrons, and photons with the nuclei of some target material. The only application in sight is an intellectual one. Thoughtful people want an understanding of the way in which nature operates. We want it in this generation—not a thousand years hence. Far too long have we groped in darkness. It has sometimes been remarked in recent years that the number of recognized particles has increased greatly since the early 1930's. Nevertheless, these particles fall into well-defined families,

Fig. 13. Rust-resistant wheat, produced by a mutation due to radiation, in comparison with the unrradiated parent stock.

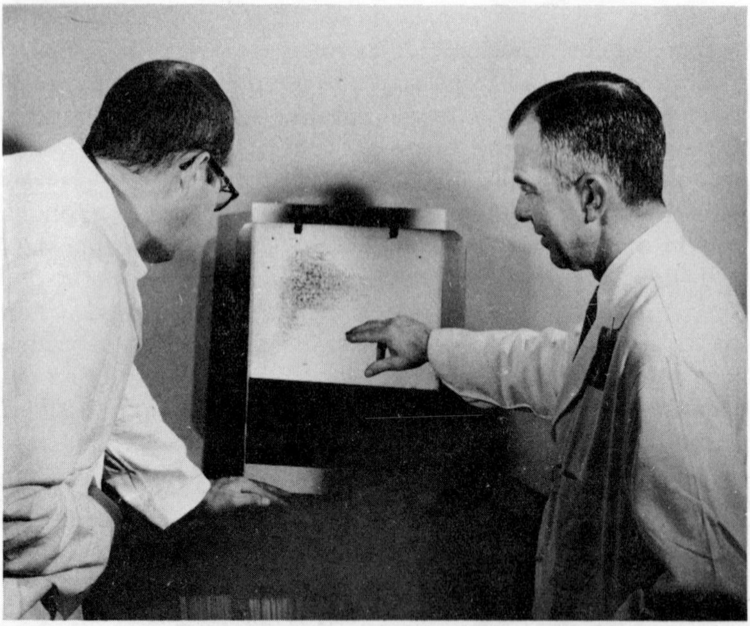

Fig. 14. Liver scan accomplished with iodine-131, revealing concentrations of the isotope, as picked up on collimated scintillation counters. The technique is basically due to P. R. Bell and his associates at Oak Ridge National Laboratory.

orderly schemes. There is a surprising amount of underlying order and simplicity in nature.

Track Photographs

A typical product of the high voltage accelerator is a set of tracks of charged particles. In the field of very high energies the tracks are customarily formed in a bubble chamber, that is, a chamber filled with suitable liquid almost ready to boil. The high energy particle makes a chain of bubbles. Figure 15 shows a track in such a chamber, with a strong magnetic field applied perpendicular to the plane of the picture. The particle travels ringwise under the influence of this field, slowing down as it goes and therefore moving on a path of decreasing radius.

The Advance in Voltage and Current

Figure 16 shows the voltages (or better, particle energies in units of a billion electron volts) for a number of recent accelerators and some projected ones. Figure 17 is a view of the installation called AGS at Brookhaven, in which the particles move on an accelerator track several hundred yards in diameter. This machine can deliver protons at energies up to 30 billion electron volts.

The Very, Very Small

Why, now, do we have this mass movement of physicists toward higher and higher voltages? The point is that the higher the voltages we use, the smaller are the details which can be examined. Figure 18 shows the order of magnitude of objects which can be studied with a variety of probing tools; on the left, a living cell visible in the optical micro-

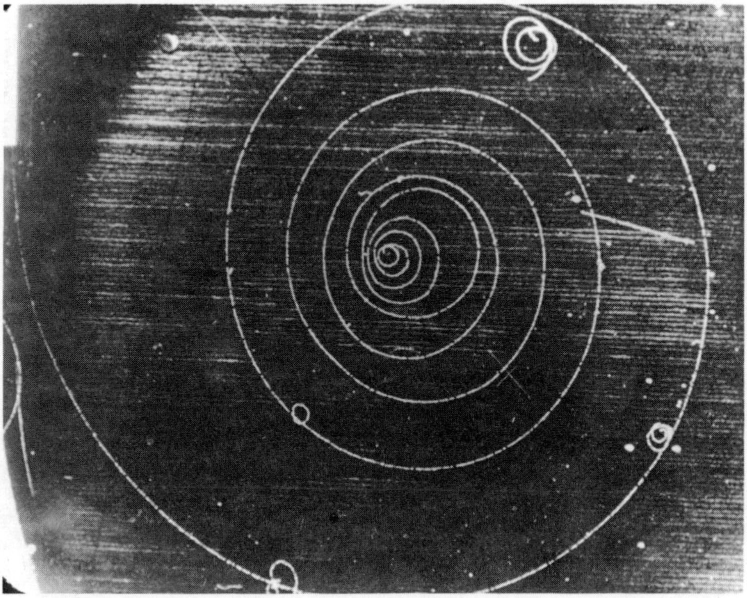

Fig. 15. Spiral track of a charged particle from a high-energy accelerator, as recorded in a bubble chamber operating in a strong magnetic field.

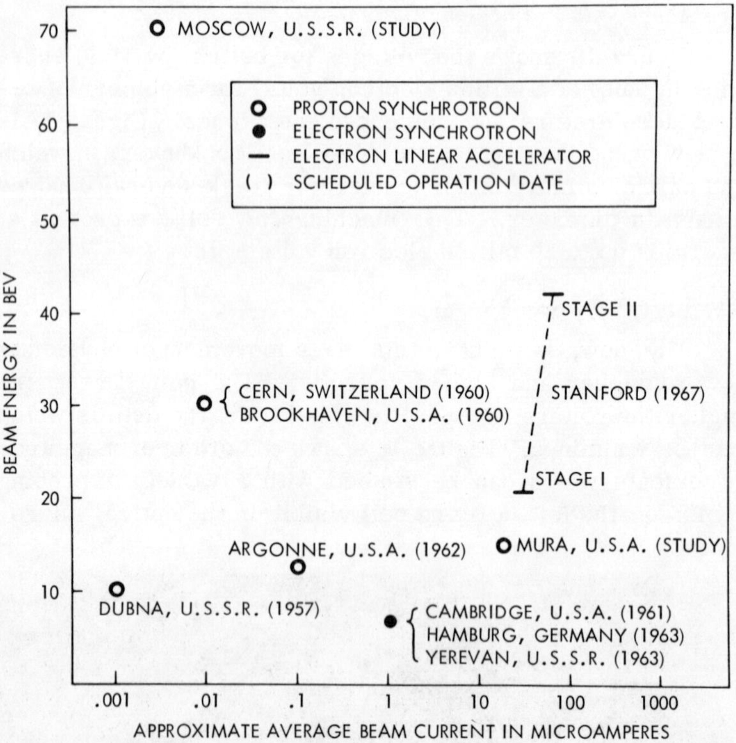

Fig. 16. Voltages and currents of high-energy accelerators. The figures give a vivid impression of advances.

scope; next, viruses at a size level of one millionth of a centimeter, visible in the electron microscope. With accelerators, we can deal with objects of an entirely different order of magnitude. Many present accelerators show the structure of the atomic nucleus at a size level of the order of 10^{-12} centimeters. For several years, physicists at Stanford University have been probing the structure of the proton and the neutron at a size level more than ten times smaller[10], and soon a new accelerator two miles long, now under construction at Stanford, will extend this range of ex-

[10] R. Hofstader, F. Bumiller, and M. R. Yearian, Rev. Mod. Phys., 30, (1958), 482.

aminable structures down to 10^{-14} or 10^{-15} centimeters. This is the only way we know to do the job.

Nuclear Reactions in Astrophysics

Now let us turn from the very small to the very large. Light and radio waves have been the standard tools of the astronomer for studying not only the stars of our galaxy but also the conditions in distant galaxies. But the light coming from the stars, all the glory of the nighttime sky, is caused by the thermonuclear reactions deep within these bodies.

Sometimes a star gets out of control and then we have a big explosion—a nova or a supernova. In recent years Lynds and Sandage[11] have detected a much more violent

[11] C. R. Lynds and A. R. Sandage, Astrophys. J., 137, (1963), 1005.
C. R. Burbidge, E. M. Burbidge, and A. R. Sandage, Rev. Mod. Phys., 35, (1947), 947.

Fig. 17. An airview of the AGS accelerator facility at Brookhaven. AGS means alternating gradient accelerator.

	OPTICAL MICROSCOPE	ELECTRON MICROSCOPE	STANFORD MARK III ACCELERATOR	PROPOSED TWO-MILE ACCELERATOR
"Light" used	Visible light	Electrons	Electrons	Electrons
What is seen	Image	Image	Diffraction pattern	Diffraction pattern
Particle or photon energy	2 ev	50,000 ev	700 Mev	40 Bev
Wavelength	10^{-5} cm	10^{-9} cm	10^{-13} cm	10^{-14} to 10^{-15} cm
Smallest object that can be "seen"	Millions of atoms	Large molecule (thousands of atoms)	Nuclei, some details of individual nucleons	Details of elementary particles
Typical objects of study	10^{-3} cm Living Cell	10^{-6} cm Virus	10^{-12} cm Atomic Nucleus	10^{-13} cm Proton

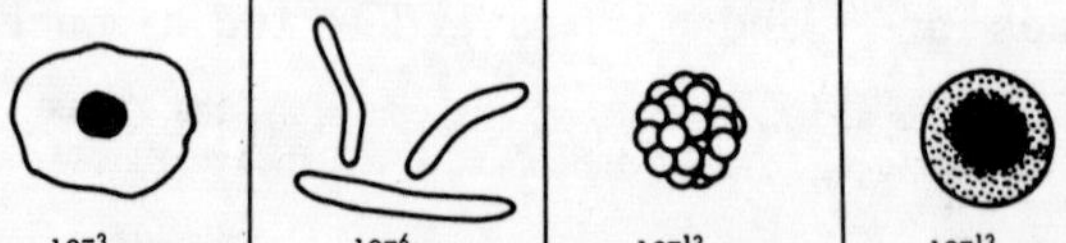

Fig. 18. Indicating the size-detecting capabilities of several types of physical equipment. From Neal, in "75th Anniversary Symposium on Engineering for Major Scientific Programs," Georgia Institute of Technology.

explosion in the center of a very distant galaxy called M-82. This striking event is shown in Fig. 19. According to one interpretation, the explosion is believed to involve about five million stars. The material coming out from the center is still expanding at speeds characteristic of the products from thermonuclear reactions. Apparently, the stars were triggered off, one after another, when they got too close together in the center of that galaxy. It is intriguing to note that if such an explosion was initiated at the center of our own galaxy five thousand years ago it might still be centuries before we received the first signal of radiation from such an event, because the bulk of evidence indicates that we are well out on the edge of a galaxy whose major radius is of the order of ten thousand light years.

These galaxies are like lamps set out in space to illuminate our path toward better understanding of the universe. They help us to see that atoms at almost inconceivable distances behave in the same way as those close at hand. Since the light which comes to us from some of these distant objects left them billions of years ago, we can appreciate that the laws of nature have great permanence. Such observations provide a firm basis for faith in order, simplicity, and a decent sort of temporal uniformity in nature.

BY WAY OF EPILOGUE[12]

It is fine to be living in a time when our knowledge of the very large and the very small and the very complex is expanding so rapidly. The nuclear studies we have considered in this brief hour are a demonstration of the basic

[12] A general discussion of the field is in: D. J. Hughes, On Nuclear Energy, Harvard University Press, (Cambridge: 1957), 258 pp.

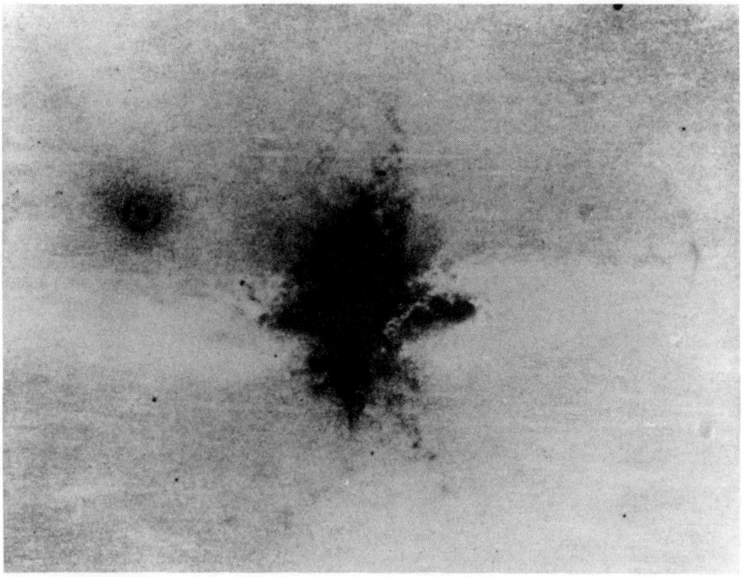

Fig. 19. Explosion in the galaxy M-82, photographed with a filter passing only the red light of hydrogen. Presumably due to grand-scale release of nuclear energy from a multitude of stars, ejecting material on both sides of the galactic plane. From Burbidge, Burbidge, and Sandage, Rev. Mod. Phys. 35:954 (October 1963).

unity of all fields of exact knowledge. Our science and our philosophy are automatically intertwined. Within our lifetimes, many of the leading problems now in sight will be solved, and other important ones will come into view. In the words of Dr. Isaiah Bowman, the late president of The Johns Hopkins University, "Scientific work has endless horizons."

ACKNOWLEDGMENTS

For aid in securing illustrations for this article, I wish to thank my colleagues, Mr. Richard Hamburger, Mrs. Kathryn E. Lockridge, Mr. Elton Lord, Dr. Robert D. Nininger, Dr. George J. Rotariu, and Miss Dorothy Smith.

Population Trends and Population Control

Ansley J. Coale

***Ansley J. Coale** is Professor of Economics, William Church Osborn Professor of Public Affairs, and Director of the Office of Population Research at Princeton University, where he has taught since 1947. He is a member of the Advisory Committee on Population of the Bureau of the Census and the Committee on Population of the National Academy of Science. He serves as the United States Representative to the Population Commission of the United Nations. Professor Coale's research on population has emphasized the economic implications of population growth, the factors determining the rate of increase and the age structure of populations, and methods of estimating past and future demographic trends.*

Population Trends and Population Control

THE POPULATION TRENDS most discussed nowadays are trends in the rate of population increase. This discussion is no exception. Banal as the subject may be, I am going to talk about trends in mortality and fertility, and in the rate of population increase. I shall begin with the trends in the industrialized areas of the world, and then turn to the underdeveloped countries. A separate description is warranted by the difference in the trends themselves and by the different implications of population trends in rich and poor countries.

In almost all of the industrialized countries—countries with relative high incomes, where the population is at least 40% urban, where agriculture is no longer the dominant occupation, where 80% or more of the young adults are literate, where modern rather than traditional methods of production dominate—the average age at which people die has increased to about 70 years. The fall in the death rate occurred gradually in the countries where the decline began long ago, as in Scandinavia, and occurred much more rapidly in countries such as Italy, Japan, and the Soviet Union, where death rates were still high at the beginning of this century. The achievement of an average duration of life of threescore years and ten can rightly be considered one of the proudest successes of modern science and technology. The avoidance of pain, suffering, and grief from premature mortal illness, that we enjoy today, is often forgotten in romantic recollections of the "good old days."

Industrialized countries have all shared another population trend; all have reduced fertility by 50% or more since the preindustrial phase of their history. The timing of this reduction varies, as was true of the reduction in mortality. In some countries the fall of the birth rate began more than a century ago, for example, in France and in the United States, while in others—such as Japan and the Soviet Union—most of the drop has occurred in the past two or three decades.

Recent trends—in the past 25 years—in the birth rates of the industrialized countries are not uniform. In many, a low point in the birth rate was reached just before World War II, followed by a recovery during or after the war, and by fluctuations different from country to country during the 1950's. In those countries that had higher birth rates before World War II, the postwar period has been one of sharp decline. Within the common pattern of much lower rates today than at an earlier economic stage, the industrialized areas have different current levels of fertility as well as divergent recent trends. The family size that current birth rates would bring varies from a little over four children in New Zealand to fewer than two in Japan or Hungary. Thus, all of the economically advanced, higher income countries of the world share the experience of having reduced both birth and death rates; but current mortality is much more uniform from country to country than current fertility.

There is a very conspicuous and important trend in the underdeveloped areas: the dramatic fall in death rates that has occurred recently or that is now occurring in so many countries, even though they remain primarily agricultural, still utilize traditional methods of production, and still have low average incomes. The fall in the death rate in many such countries has occurred at a more rapid pace than any decline in mortality recorded in the histories of the industrialized countries. In Ceylon, the death rate was cut in half in six years. In several countries where the estimated average duration of life in 1920 was no more than 30 or 40 years, life expectancy now exceeds 60 years: in Ceylon, Taiwan, Hong Kong, and Mexico, for example. Sometimes startling declines in death rates have occurred in the absence of any substantial changes in the backward economy. Rapid reduction in mortality has been made possible by new techniques in medicine and public health developed in the last 25 or 30 years. These range from insecticides and antibiotics to superior ways of organizing public health administration and education at low cost.

Because of these innovations and the availability of international assistance in public health, it seems likely that in those countries where death rates remain high, a rapid and extensive decline in mortality is an early prospect. Present differences in the death rates of underdeveloped countries are still large because all have not yet shared effectively in the possibilities of low-cost modern health techniques, but it appears probable that in the absence of war or major political disorder, these differences in mortality should narrow, and in the next generation most large populations should come to share an expectation of life of 60 years or more.

The trend in fertility in the underdeveloped countries is nearly uniform—a birth rate essentially unchanged at a high level. There are a few exceptions. The Chinese populations outside of the mainland, in Singapore, Taiwan, and Hong Kong, have sharply reduced their birth rate in the last five or six years. In one or two other instances, recorded birth rates have fallen, but because of the uncertain quality of the statistics, we cannot be sure of existence of a genuine trend.

What trends in growth do these birth and death trends imply? The growth in the world population has, except for transient variations, steadily accelerated during the last two or three centuries. Prior to 1650, the average rate of increase of mankind was very slow, so slow that more than a thousand years on the average was required for the population to double. After 1650, however, the population of the world doubled in less than 200 years. The next doubling, between 1840 and 1930, took less than a century. The current rate of increase would bring a doubling every 35 years. This continued acceleration that recently has become very sharp is the result of a tendency for the global death rate to go down, while the world birth rate has remained nearly constant. The industrialized areas, constituting only a minority of the world's population, contributed to the early acceleration because the decline in mortality in these areas was earlier on the average than the decline in fertility. In the underdeveloped areas, there was probably a gradual and not very extensive fall in mortality before the sharp decline in the mid-twentieth century. The recent and current rapid acceleration in world population growth is primarily the result of the fall in the death rate in the underdeveloped areas, buttressed to some degree by the recovery in fertility in the most highly industrialized countries from a low point just before World War II.

The growth of the industrialized countries themselves today is a mixture of slow to moderate increase. Differences in growth rates among the developed countries are dominated by fertility differences. In fact, the substitution of the risks of death at each age in one industrialized country for the risks prevalent in another would have only a very small effect on the rate of increase. Current rates of increase in the industrialized countries would lead to a doubling only every hundred years or so for some of the countries of Western Europe, and to a doubling every 35 to 40 years in the most rapidly growing populations. The fastest growth in the industrialized countries barely equals the world average.

In contrast, in the low-income countries population growth ranges from moderate to very rapid. Generally speaking, where the current growth rate of underdeveloped areas is moderate, it is accelerating sharply. Among these populations, growth differs because of differences in the level both of fertility and mortality; but as the international promotion of modern public health methods proceeds, differences in growth rate arising from mortality differences will tend to disappear. The duration of further accelerated growth in the underdeveloped world as a whole and the maximum growth attained will depend upon the speed with which effective health techniques can be spread, and upon whether the reduction of fertility becomes widespread in the next few decades or remains confined to two or three areas.

In sum, then, the much publicized population explosion in the world is primarily an explosion of the population of underdeveloped countries, and has been caused by a precipitous fall in death rates accompanied by little or no change in birth rates.

EFFECTS OF POPULATION TRENDS

The implications of these population trends are wholly different in the industrialized countries on the one hand and in the underdeveloped areas on the other. The slow-to-moderate growth of the population of industrialized countries poses no major economic difficulties, at least in the near future. There seem to be no insuperable economic barriers to supporting for a century or so at ever rising standards of living the greater numbers that continued growth would bring in the most advanced economies. The United States has one of the most rapidly growing populations among industrialized nations. A continuation of current fertility and mortality would produce a population of about a billion in a century. The esthetic, psychological, and political consequences of such further crowding are doubtless unfavorable, but the prospect of continued technical progress, and the absence of deprivation in crowded Europe make it appear probable that even these numbers could be supported at a substantially higher level of living than today.

In the long run, even the wealthiest country cannot

tolerate a continuation of a constant positive percent rate of increase, because such continued increase leads in the long run to overwhelming numbers—to such absurd situations as the density of one person per square foot that would result in the United States from continued growth for about six or seven hundred years.

In the underdeveloped countries, on the other hand, the disastrous consequences of continued population growth are not so distant, because the rate of increase in the underdeveloped countries is much greater and still accelerating, and because these countries are less able to afford a rapidly growing population. Mexico illustrates the implications of more rapid increase, *per se*. Its current growth rate would multiply the population by 32 in a century, compared with our prospective increase by a factor of a little more than five in a century. Mexico today has a population of 39 million compared with 190 million in the United States, yet it would reach a billion, as we would, in about a century. Mexico would then grow to 32 billion a century later.

More immediately, high fertility and a high rate of increase are impeding the economic development of the low-income countries right now. National output must grow rapidly just to keep up with the increase in population. It requires new factories, highways, irrigation works, and the like if a country is to modernize and raise its national product rapidly. A high birth rate and a rapidly growing population require that resources be devoted to producing goods for immediate consumption and housing rather than to building new factories. Actually, a high birth rate reduces the proportion of the national effort that can be devoted to expanding industrial capacity. In fact, for about 30 years, a country that reduces its high birth rate can increase its total national product faster than one that keeps its high birth rate unchanged—in addition to the advantage that a low birth rate country has of dividing its product among fewer persons.

Industrialized countries can certainly support the populations resulting from their slow or moderate rates of growth at least for the next generation or two, and at the same time provide rising living standards for their populations. However, the rapid and accelerating growth of the populations of underdeveloped countries is making the escape from poverty in these areas slower and less certain. Both kinds of countries face the long run prospect of a catastrophic increase in density, but a much more immediate prospect for the low-income countries.

THE CONTROL OF POPULATION GROWTH

Throughout man's experience, population size and population growth have been regulated to a large degree by variations in the death rate—by the classic positive checks of Malthus of disease, war, and famine. However, in the peacetime experience of modern industrial countries, variations in the death rate are negligible factors in the rate of increase. This negligible effect of mortality (so long as fertility is above replacement) results from the fact that survival to the ages of parenthood among women is already very near to 100% (98% in Sweden, for example). Further reductions in mortality would doubtless keep more old persons alive, but would have hardly any effect on future births, the source of future population increase. In other words, neither current differences in mortality rates in the industrialized countries, nor further reductions in the future play any significant role in regulating the long-run rate of increase of their populations. In the underdeveloped areas, a similar situation seems to be emerging, and given continued progress in the spread of public health, it will prevail by the end of the century.

Few would welcome a return to variable mortality as the regulator of growth. But growth is inevitably regulated. Indeed, the average rate of increase over very long periods of time must be close to zero. At a 2% rate of increase, starting from today, it would take only about 1200 years for the human population to outweigh the earth. At 1%

2400 years would be required, and at one-half of 1% about 4800 years. In short, the continuation of even moderate rate of increase for a few centuries would lead to densities of population that could not possibly be sustained, no matter what miracles science performs. Moreover, in the underdeveloped areas the interval during which sustenance can be found for a population growing at current rates is measured at most in generations, rather than in centuries. If population growth does not come to be restrained by a reduction in fertility, mortality must resume its regulative role.

THE CONTROL OF FERTILITY

What forces determine the level of human fertility? There are, first of all, biological limits to human fertility. The capacity to bear children is confined to an age interval of approximately 15 to 45 years in the human female. The maximum number of births per woman that could be achieved on the average in any large population (unless specially selected for unusually high fecundity) is perhaps 13 to 15, but in no society of which there are accurate records, is any such high level of fertility attained. The maximum on record is eight to ten. In preindustrial populations, fertility is kept well below the biologically possible maximum by various customs, taboos, and sanctions. For example, there are customs or laws governing the proportion of women of childbearing age who are partners in a marriage or other form of fertile union—customs governing the incidence of spinsterhood, the age of marriage, the prevalence of divorce and widowhood, the possibility of remarriage, and the like. Moreover, fertility of the married can be affected by habitual usage with respect to periods of separation, taboos on intercourse (for instance during lactation), the duration of nursing, etc. What appears to be rare in preindustrial societies (though not unknown) is the control of fertility through conscious, deliberate methods, making it possible for married couples to decide on the number of children they shall have.

The control of fertility within marriage is characteristic of the economically advanced countries. Married couples in these countries behave in ways deliberately designed to prevent conception. In some countries (Japan and in Eastern Europe) many births have also been prevented by induced abortion.

So far as we can judge today, the modern decline in fertility started first in France, before the end of the eighteenth century. There was also a very early decline in the United States (beginning no later than 1800), but in the early years fertility decline in our country may have been caused primarily by later marriage and increased spinsterhood rather than the limitation of births within marriage. In Europe outside of France, the decline in fertility began after the middle of the nineteenth century. It has now become universal in Europe with the exception of Albania and one or two isolated provinces in the Balkans. It is clear that this reduction in the birth rate which in nearly all of today's highly industrialized countries has been at least 50%, was the result of the extension through the population of the voluntary control of fertility within marriage. The sole notable exception is Ireland, where martial fertility has remained very high, and the birth rate has been reduced by the postponement and avoidance of marriage. The average age of marriage for women is nearly 30, and almost 25% remain unmarried at age 45.

The fundamental difference between fertility in a preindustrial and an industrialized country is that in the underdeveloped area, the birth rate is controlled by customs which are for the most part accepted without question, and in the industrialized countries, fertility is controlled by techniques that make each birth the result of a deliberate choice. To oversimplify slightly, one might say that in underdeveloped countries births are something which parents accept as natural, like the weather, whereas in industrialized countries births occur because the parents consciously make a decision, as when they buy a new car.

In the United States today, the voluntary control of fertility is very widespread. According to responses in nationwide surveys, more than 90% of white couples who suffer from no impediment to fecundity, at some time during their marriage practice some form of contraception (including in that term, periodic continence or the safe period). Only a small fraction of American couples, mostly the underprivileged and poorly educated, do not consciously restrict their fertility.

The level of the birth rate in a country is not necessarily an indication of the prevalence of deliberate family limitation. For example, in the United States after World War II, fertility has been about as high as it was in the first decade of this century. Before the first World War, there was in the United States a group of effective controllers of fertility who had very small families; and the remainder of the population did not exercise voluntary control and had large families. Subsequently, the practice of limitation spread through the population and by the 1930's fertility had fallen to the point where the population was barely replacing itself. However, during the recovery of the birth rate in the 1940's and the 1950's, American couples have not lost or abandoned the knowledge or practice of family limitation. There is no doubt that the number of couples knowing about and practicing birth control is higher today than in the period just before World War II, and certainly higher than before the first World War. Our current moderate fertility is a result of the fact that a very high proportion of our population prefers to get married at early age, to avoid childlessness or a one-child family and to have two, three, or four children. Bachelorhood and spinsterhood are now avoided, as are the childless family and the one-child family, but there has been no return to uncontrolled fertility, or to the high proportion of families with six, seven, and eight children characteristic of the early days of the century.

I turn now to the importance of attitudes in the development of controlled fertility. To judge from the evidence available, it was the change in attitudes towards family size and the possibility of controlling fertility, rather than new inventions or newly discovered techniques in the field of birth control, that was responsible for the first development of family limitation. It seems almost certain that the birth rate in France and indeed in most of Europe fell through the use of *coitus interruptus* rather than through mechanical or chemical methods of contraception. This is a method mentioned in the Old Testament and part of the folklore of almost every society. In other words, the control of family size occurred because people developed the idea that control was an acceptable mode of behavior and because they desired strongly enough to restrict the number of births. Even in the United States today, the strength of motivation is very important to the success of contraception, whatever the method employed. Colleagues of mine at Princeton have found in their survey research that the effectiveness of contraceptive practice for any given method is several times higher (i.e., failure rates several times lower) among couples who have had as many children as they say that they want, than it is among couples who want more children and are using contraception only for spacing. On the other hand, in underdeveloped countries, the vast majority still have children without deliberate decisions. Women who in an interview indicate a preference for three or four children only, are unable or unwilling to modify their behavior and in fact have six, seven, or eight—even though they have no moral or religious scruples governing contraception. A change in attitude from resignation to voluntary control must come before births can be regulated by conscious decisions.

In the face of the widespread apathy toward the regulation of fertility in populations where high fertility is inhibiting the success of social and economic development, scientific research has been intensified in the whole field of the physiology of human reproduction. From this research may come better techniques for detecting and predicting the onset of ovulation, so that periodic continence may be more effective. There are already emerging improved methods of contraception: pills, innoculations that might confer six months or more of temporary sterility, new mechanical devices that although apparently harmless prevent conception until they are removed, and ways of inhibiting the male as well as the female contribution to conception. The result of this research may be methods of contraception that are more certain, easier to use, less expensive, less distasteful, and requiring less of a change in attitude, or at least a lower level of motivation, for effectiveness. This research is paralleled by attempts to develop new techniques of education, communication, and persuasion to overcome apathy and resignation. In other words, new methods of contraception may reduce the level of awareness and concern required to limit families, and educational campaigns may arouse the requisite awareness, so that the number of children comes to be a matter of conscious decision in the low-income countries as well as in their richer contemporaries.

In speaking at an outstanding Catholic university, I cannot resist remarking Catholic attitudes towards the control of fertility. It is of course Catholic doctrine that only continence—continuous continence or periodic continence, also called the safe period—is a licit means of controlling fertility. It is not my place—I am not a Catholic, a theologian, or a moral philosopher—to express any opinion about doctrine. However, I would like to note the apparent existence among Catholics in the United States, and some other countries, of a pronatalist attitude—an attitude in favor of large families as such. When questioned in a nationwide survey of urban women who had just borne a second child, Catholic wives said they wanted 3.6 children on the average, while non-Catholics wanted fewer than 3. The more significant finding in this connection is that the more education women had received within Catholic institutions, the larger the families they wanted. Those whose schooling had been wholly from Church institutions and who had attended college, wanted an average of 5.1 children. I wonder what is the source and the basis of this attitude in favor of many children among those who apparently have been most exposed to Catholic indoctrination. Professor Westoff at Princeton is now studying the change in attitudes in college among Catholic and non-Catholic girls, in an effort to find a partial answer.

I do not believe that an uncritical approval of a high birth rate *per se* has any doctrinal basis. Support for high fertility and unyielding opposition to fertility control would, if effective, ultimately and inevitably bring back high death rates. Opposition to the control of fertility impedes (even in the short run) the attainment of universal education, adequate housing and food, productive employment—indeed the attainment of a decent and dignified life in the underdeveloped areas.

The Impact of New Materials and New Instrumentation on our Foreseeable Technology

E. R. Piore

Emanuel R. Piore *is vice president and chief scientist, and a member of the board of directors of International Business Machines Corporation. He joined IBM in 1956 as director of research. Dr. Piore was associated with the Office of Naval Research from 1946 to 1955, serving as chief scientist for the last four years of this period. Prior to joining IBM, he was Vice President for Research of the Avco Corporation. He is a member of the National Science Board, the Defense Industry Advisory Council, and a former member of the President's Science Advisory Committee. Other memberships include the board of Science Research Associates, Inc., and the corporation of Polytechnic Institute of Brooklyn. He is a trustee of the Sloan-Kettering Institute for Cancer Research and the Woods Hole Oceanographic Institution. He is a member of the National Academy of Sciences, and a fellow of the American Physical Society, the Institute of Electrical and Electronics Engineers, the American Academy of Arts and Sciences, and the American Association for the Advancement of Science. Dr. Piore received his A.B. and Ph.D. degrees in physics from the University of Wisconsin in 1930 and 1935. He also served as instructor at Wisconsin from 1930 to 1935. In 1962 he received the honorary degree of Doctor of Science from Union College.*

The Impact of New Materials and New Instrumentation on our Foreseeable Technology

BEFORE LOOKING INTO the future of technology, I must spend a few moments clarifying some concepts. I would like to differentiate between science and technology. Although they are related, they do not have a common history or roots.

Technology is meaningful only if it finds a use in our society. We can recognize the technological developments only when they have had an impact on the way we live, or do things, or transact trade. This may be in the area of health; it may be in the area of manufacturing; it may be in the area of warfare; it may be in the area of economic revolution. Technological revolutions are recognized when a technology has been used, has been widely accepted, and, if I may be very crude, when someone has made some money out of it.

This is in contrast to science, which is a body of knowledge. Every scientific field concerns itself with certain intellectual structures that the human mind has created to fit the bits and pieces of scientific information and knowledge into a greater understanding of the world around us. It can be observed that scientific growth and economic growth have occurred concurrently and there is an element of technology coupling them. It has been possible only recently to conduct any analysis that demonstrates that there is some kind of relation between science, technology, and economic growth. I know this may not be acceptable to many of you but this is my view. I would like for a moment to inject some history to indicate my thesis; I would like to deal with three revolutions. What is important historically is often dramatized by calling it a revolution. These revolutions have shaped our contemporary society.

The first revolution is known as the Renaissance and is usually associated with humanism—the arts and the rediscovery of the Greek writings. But I think from our point of view I would like to be a bit vulgar and say the importance of the Renaissance was a revolution in trade. Expansion of trade, expansion of communication, expansion of the movement of people were basically part of the Renaissance. The other two revolutions are the scientific and the technological revolutions. These three revolutions occurred concurrently but did not have common intellectual roots.

If one looks at the technological revolution one observes that many things were invented, many devices were brought into being, many materials were made available for use. They became part and parcel of our physical civilization. Yet this occurred quite independently of the scientific revolution.

The men responsible for these inventions and innovations, their exploitation, and for the new concepts of science belonged to three different groups. By the time the industrial revolution appeared on the scene, those working in technology and those whom we call entrepreneurs, in the economic textbook sense, started to be coupled together. The coupling between science and technology is a very contemporary phenomenon and becomes visible to an observer in the nineteenth century. At present this interaction is very intense. Let me cite some examples.

Steam was used as a source of power to pump water out of mines almost as far back as the Middle Ages, before there was any understanding of thermodynamics. The rewards and recognition in science were, and are, quite different from the rewards and recognition in technology. Science deals with great humanistic and cultural values and can stand quite independently as a human activity, although it is related in the contemporary culture to economic growth and technological development. Some of these relationships I will explore.

If we try to look at what technology holds for the future and try to make predictions about the effect technology will have on our daily lives, the way we live with each other, the way we discourse with each other, we have to look at a number of places in order to get some indication for a basis of these predictions. Some of these places are the scientific and engineering laboratories in our society. We have to understand the problems the scientists are tackling, the instruments they are using, the techniques they are evolving. If that had been done in the eighteenth century, no valid predictions could have been made since there had been no historical experiences. But these days we have this experience and can make some judgements. Still another place to look is the government and how it is spending its money. If the government decides to spend a great deal of money on oceanography, high energy physics, or space, this affects our technological future. In addition, there is something that is very difficult to describe and that is the impact of inventions. Let me cite an example or two.

When Alexander Graham Bell worked on the telephone, it was not known whether this was a great technological innovation in the sense that it would appear in the market place and change our way of life, or whether it was a toy. This was impossible to predict. But now you know very well that the telephone is a source of sorrow and joy and many other things. It violates our privacy to an ever-increasing extent.

In contrast, there is an invention that never got off the ground. You will have seen beautiful etchings and drawings of ladies with parasols watching hot air balloons over Paris. The balloon has had no impact on our way of life and I do not consider its invention an important technological event. Presently we use very sophisticated plastics as balloon material and get scientific information, but as far as our daily lives are concerned it does not make very much difference. But the balloon is an important scientific instrument.

Another important area of investigation is technology as it is independent of science. We look at things that people are using and consider self-generated technological dynamics, those activities that will provide improvement even if science lacks wisdom in the area. I am going to select a computer as a case study, focusing on it as something that exists in our society, and indicate how its technological development came about. Then I am going to come back and give you further indication of where I think our instruments and materials will take us in terms of low energy, high energy, extremes of temperature, and extremes of pressure. In trying to predict the future, we must look at activities in the laboratories at the extremes in technological developments. In order to do this, let me give you very quickly one or two aspects of the computer. When we talk about contemporary computation, I think we first must indicate the important part John Von Neumann, one of the great intellects of our time, played in this technological development. He convinced the scientific community that we could not make further progress in science without developing this instrument, the computer. Never mind that other people had made great inventions and never mind the long history of thinking about mechanical computation, John Von Neumann was the one who convinced the American intellectual community that this was the way to go.

In order to talk about the computer, bring into focus the importance of low energies, and discuss what is being done with technology, let me indicate very briefly that the computer broadly has three components: an arithmetic unit that does arithmetic, a unit that stores information and input, and output units that permit a human being to use the computer and obtain solutions to important problems. Another unique property of the computer is that it is a digital device and operates in binary form. This is a very important aspect because it is going to characterize how people will read and write in the future and characterize our technological world in the future. When we say the computer is digital or binary and performs only one operation, we mean that it says either yes or no, zero or one, to each question asked. If you take 18-year-olds with any aptitude in mathematics, put them in a computer installation and give them a piece of paper full of holes, they can read the holes in very much the same manner as they read the printed page. It is a natural language to them. This most likely is the most important aspect of our technological revolution.

Let me go back to the computer, as it characterizes much of our technological activity these days. To evoke this yes and no decision, we look for changes in the state of material that occur with very little energy. We are looking for changes that can be observed visually. This is not a question of changing from a solid to a liquid, but basically a quantum-mechanical operation.

Until recently we made these small energy changes with comparatively large uses of power, as in vacuum tubes. More recently, we have used transistors, large transistors, the size of, say, a small fingernail. Now we have reduced the size, and thus the power required, so that you can get about a thousand of these devices into a circle of one inch in diameter. We have acquired the ability to obtain measurable changes of state with smaller and smaller amounts of power.

Another problem was how to operate the device above the noise level. The signal must be above the random processes that go on in nature. It has been demonstrated—and this involves the physicist, the chemist, and the metallurgist—that presently very few electrons are required to detect the signal above noise. In the most extreme case, and this is an old experiment, it takes about 16 quanta for an eye to detect the signals. Most of these quanta are actually lost in the absorption process of the eye fluids and in reflections from the surfaces of the eye structure.

Thus we have reached a point in technology where we can get signals from smaller and smaller bits of matter, and the problem becomes one of art, engineering, and invention to get in and out of these small bits of matter. The shrinking of the size of the transistor makes new tools available for biology and medicine, and these will be the tools that will assist revolutions in these fields.

The other important part of a computer is its memory element. Again, this is a problem of changing the state of matter, using very little material and expending very little power. As the size of the memory element shrinks, the total number of elements in the memory array can become larger and larger. If you are interested in this general topic, I suggest that you look into the work of a theoretical physicist, Richard Feynman, of the California Institute of Technology, who wrote a very interesting paper that appeared in the "Cal Tech Engineering and Science Magazine" in February, 1960, where among many other things he demonstrates that you can put the whole 24 volumes of the "Encyclopedia Britannica" on the head of a pin. The basic problem of making this type of information storage useful is how to get at the information and get it out. What I am predicting is that we are going in that direction and that we will solve the problem.

Once we are in this area of small energy and low noise level, the obvious thing to do is go to a very low temperature, in the neighborhood of absolute zero. The possibility of operations at a low temperature is a good example of the productive relationship between science and technology. Physicists started to tackle this problem 50 to 60 years ago but we have yet to have a technological development out of it, although it is beginning to appear. The advantage of low temperatures close to absolute zero is that at these temperatures changing states requires very little power, superconductivity occurs, and ambient noise is reduced. An example is the solid state detecting devices operating at liquid helium temperature which we use to receive signals from communication satellites. Ultimately we are going to operate most of our important electronic devices at liquid helium temperatures.

Matthias, of Bell Laboratories, now at the University of California at San Diego, has demonstrated that there are many materials and compounds that are superconductors. To date, the only thing we have done with superconduction in the technological sense is to build superconducting electromagnets capable of very intense magnetic fields, on the order of 100,000 gauss.

The most interesting part of all this, and General Electric is studying it, may be that an economical way to transmit electric power will be developed and used, making possible very high current, no resistance, and no power losses. Ultimately we will have transmission lines carrying power throughout the country in helium baths. This is possible and will be done, even though there are those who disagree. Their attitudes are comparable to those of the people who never believed that nuclear power would compete with fused fossil power and are now being proven wrong. Superconductivity has yet to be made into a technology, but quite apart from transmitting power, there is every indication that it may become a very important element in computing machines.

I have mentioned that in a very large measure the government determines the direction of technology. It does this in two ways: first, by putting impossible requirements on our technological resources and, secondly, by actually supporting certain fields of science. It may not be realized that we would never have had a transistor, we would never have had superconducting magnets, we would never have shrunk the size of the computer or the size of its memory, without the pressure and the money that the government supplied in accelerating the development of the transistor and the use of these miniature devices. Such pressures will continue and are almost more important than the actual events occurring in science. I know this may be heresy, but these are the realities of the present world. I have already stated that superconductivity as a scientific problem goes back 50 to 60 years but it is only under government pressure that we will get some type of technological development. The transistor is the best example of this. Although all the commercial companies are going into what is popularly known as microminiaturization, this would not have occurred without government funds.

This characterizes the world we live in. If you say the wonders of science will create the wonders of technology, you are underestimating the creative ability of engineers and technicians when they have to work under forced draft, even when they do not fully understand the character of the external world. We get so interested in basic research and its challenge, we lose sight of the fact that at times some things are done and understanding is acquired later. The principal force behind getting things done and understanding them later is the government, not the itinerant inventor.

The laser is another good example of government intervention. Though I am told it will be used to track the Explorer Satellite, it is one of the great inventions of our contemporary world that is still lacking application. It has not as yet been made into a technology although we know how to modulate its frequency. At this time the most interesting use of the laser is to burn holes in stainless steel razor blades. There are many ways of doing that.

If you want to get a flavor of the pressure that the government applies on a future technology, I suggest you read an article on future scientific spacecraft in the April, 1964, issue of "Physics Today."

I have used the computer as a focal point of something in the market place that generates its own growth and indicates that technology at times runs independent of science and has its own inner soul, just as science has an inner soul, its own goals, and its own great intellectual accomplishments. If you go into any laboratory these days, and I say any laboratory, you will find that the computer is used very broadly. You will also find, for example, in the high energy physics laboratories at Berkeley, Brookhaven, or Geneva that you cannot observe the strange events you are looking for directly but must observe them with instruments and photographic plates. The computer is instructed to identify strange events on statistical or other bases. This is true also in looking at biological materials with X ray. You look at a sample and the final result you see is the print-out of the computer. This characterizes contemporary scientific research. At Cal Tech they are working with a poor little wiggling fly; in trying to understand its energy consumption and all its motions, data are fed directly into a computer.

A computer is a digital device, a binary device. Most of our instruments to date have been analog devices that can easily translate whatever information we require into a form that a human can read and understand. A human does not read digital information. However, all our future instruments, and I am making an almost absolute statement, will be digital in form and digital in measurement. This is a radical departure. Mathematics is not expressed in these terms. Our children, however, will have no problem with this. Scientific information will be fed directly into the computer and experimental results will be read out in the form of a print-out, on a cathode ray tube, or some other device. This technique is occurring wherever you turn. It is the future of biological research, the future of chemical research, the future of physical research.

There are a few obvious observations I would like to make, for completeness more than anything else. I have indicated that we are now beginning to use low temperatures and are perfectly at home operating close to absolute zero. Many companies are building continous refrigerators. The reason we feel so at home with low temperatures is that at one period in weapons development we needed a great deal of liquid hydrogen and had to learn to make it by the ton and transport it across the country. Now we are going to use low temperatures.

We also are beginning to feel at home with very high pressures and very high temperatures. This is in part because of government programs concerned with the nose cone reentry problem. The development of synthetic diamonds may have been related to this.

In order not to slight biology, let me quote Professor Melvin Calvin of Berkeley, a very distinguished biochemist, to indicate his point of view. As we read this we realize that biological research cannot be done without computers. The word I want you to be aware of is "system," because one of the reasons we have made great progress in biology is that we have been able to keep the system alive and make intrusions in specific places in that system. The following is quoted from Calvin:

> "Techniques for the detection of minute systematic variations in the property of any system which heretofore were considered beyond the region of our senses, either natural or constructed, are now becoming useful tools for our examination in the world around us. Previously such minute systematic changes were lost in the random, or thermal, 'noise' of our environment. Fluctuations in our environment which once appeared random are subject, with computer techniques, to closer inspection which may define precise relationships. Our methods of amplification, in other words, and signal analysis make possible investigations both of the physical and biological worlds at levels inconceivable only a decade or two ago. Methods of separation of the complex molecular mixtures of nature and the analysis of the separated entities, both in terms of their composition and detailed structure, have reached new levels. This, in turn, has permitted the synthesis, or construction, of these very complex objects, both in the inorganic and biological world, to a degree of complexity approaching that of nature itself. New regions of extremes of temperature and pressure are becoming available to us for study in the laboratory, and these also give us information about the behavior of matter under conditions not normally encountered. Some of the technological applications of these new scientific frontiers are already apparent."

Since the government has committed itself to a great program in oceanography, we are reaching a point where the old style of oceanographic research, characterized by throwing instruments over the side of a ship and pulling them out again, may be going the way of the whaling ships. There are computers on ships that can immediately process the information and in fact make it possible to perform more experiments before the cruise is over.

To summarize: Science and technology are related; their relationship grew closer and closer in the nineteenth century, and much more so in the twentieth century. Technology also has an independent life of its own that drives ahead. This drive is affected by the market place. In contemporary America, which in the technological sense is practically the world, this drive is generated by the government and its needs. Very often science comes along later to elucidate the technological accomplishment rather than lead the technological development.

This is the real world and when we talk about science and economic growth, we should be cautious in trying to make a great sermon out of it because we can be proven wrong in many instances.

Design for a Brain

Philip M. Morse

***Philip M. Morse** is Professor of Physics, Director of the Operations Research Center, and Director of the Computation Center at M.I.T. He has been Director of the Operations Research Group, O.S.R.D., and was the first Director of the Brookhaven National Laboratory. He was the founder and first President of the Operations Research Society of America and has been Executive Secretary of the International Federation of Operational Research Societies. He is Chairman of the Advisory Group on Operations Research to the Organization for Economic Co-Operation and Development, and has been Chairman of a similar panel for the Office of the Science Advisor to NATO. He is a Fellow of the National Academy of Sciences, and the American Physical Society, and the American Academy of Arts and Sciences. He is co-founder and Editor of* Annals of Physics. *His books include "Methods of Operations Research," "Queues, Inventories and Maintenance," "Vibration and Sound," "Methods of Theoretical Physics," and "Thermal Physics."*

Design for a Brain

WE ARE ALL TOO AWARE of the results of the mechanical revolution: the abolition of slavery; the vanishing of the family horse; the ability to breakfast in Paris and lunch in Washington; the traffic jams in every large city in the world; the presence of the bulldozer, which can remove all the trees from a residential site in a morning and which can regrade the site in accordance with our desires in an afternoon—to mention a few of these results. The electronic computer promises to be the instrumentality of the next stage in this revolution. Its capability as a mental assistant is largely unexplored as yet; most of us know too little about it to appreciate its potentialities.

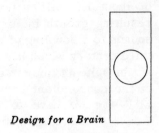

Its capabilities have grown rapidly and, to a large extent, unnoticed, even in universities, where it was first developed. For example, when I joined the faculty at MIT in 1931 there were some 20 desk calculators plus a few specialized analog computers on the campus. Today at MIT there are available to students and faculty two IBM 7094 computers, one 709, and several dozen other machines of progressively lesser capacity, exclusive of the equipment possessed by Lincoln Laboratories and other off-campus projects. Since the 7094 is, in a very crude sense, about 300,000 times faster than a desk calculator in the hands of a skilled operator, we can say that the computational capacity available today at the Institute is the equivalent of about 1.5 million desk computers plus operators. The resulting growth factor of about 60,000 in 32 years corresponds to a doubling of capacity every two years of the one-third century which has elapsed.

In spite of this fantastic growth of capacity and in spite of the concomitant growth of computer use by research at MIT—for we have to run our machines 24 hours a day, 7 days a week, to keep up with demand—we are only beginning to be aware of the potentialities of the computer. At MIT, while nearly all our students are taught computer programming, the computer is as yet little used as an aid in the educational process. Although theoretical research now depends heavily on computer assistance, our libraries still use traditional methods of keeping track of their books and periodicals. Although a few experimental research projects are programmed to have the machine automatically record and analyze measurements, we still fit student and faculty schedules to available classroom space each term by shuffling pieces of paper around. The plain truth is that we simply haven't yet learned how to use this instrumentality we have. In many ways we are at a point in development about equivalent to 1910 in the development of the automobile. The machine is being talked about by everybody, but very few have operated one; the next generation will begin to have instinctive response to the computer equivalents of brake, gearshift and steering wheel. Meantime one needs to describe the instrumentality before discussing its potentialities.

The electronic computer can be described on at least three levels: in terms of its equipment, in terms of its organization and control, and in terms of its possible uses. I cannot spend much time on the first. Present electronic computers have three parts. There is an arithmetic unit, the equivalent of the familiar desk calculator, which carries out the additions, substractions, multiplications, or divisions required by the particular program. In the faster machines the multiplication of one five-digit number by another is accomplished in a few millionths of a second.

Then there is a hierarchy of memory or storage units, in which the numbers which are to be multiplied, or have been multiplied, are stored, waiting to be processed by the arithmetic unit or waiting to be printed out as answers. These units have a variety of sizes and reaction times, ranging from the magnetic tape, which can store several million ten-digit numbers or seven-letter words but which takes several milliseconds to "recall" a particular word, through the magnetic disk or drum, with a capacity of a few hundred thousand words and reaction time about a millisecond, to the high-speed memory with present capacity of 10 to 30 thousand words and speed equal to that of the arithmetic unit. Information flows from slower to faster memory, to be processed by the arithmetic unit, then flows back down to be printed out as solution.

Finally there is the control unit, which schedules this two-way flow of data and solutions and which tells the arithmetic unit where to find and what to do with the next numbers to be processed. In the early machines this sequence of operations, this program, was accomplished by rewiring or reconnecting the various units for each new problem; in many analogue machines this method of control is still in use. Later computers used a stack of punched cards or a punched paper tape, which fed into the machine a different sequence of instructions for each different program. Nowadays the sequence of instructions is stored in the memory, along with the data and the answers. The machine then reads some words in memory as instructions, telling it what to do next, and some words as data, to be processed in accord with the instructions.

This step in development, to store the program along with the data, was first suggested by von Neumann, one of the few top-ranking mathematicians who has interested himself in the logical problems of machine organization. It opened up a new dimension of capability for the computer. For now the instructions themselves could be operated on by the machine, to be modified in the light of the progress of the program. The machine could be told to do this or do that, depending on what has happened earlier in the program.

This brings us to the second level of description of a digital computer, which has to do with the nature of the programs of operating instructions for the machine, how they control and how they are, in turn, modified by the computer. Two aspects of the new dimension made possible by von Neumann's suggestion are called the loop and the conditional. Many operations of the computer are repetitive. In summing a series, for example, each successive term is a product of several factors, each factor different from, but related to, the factors in the previous term. The machine multiplies the factors for the n th term together and adds it to the sum of the n -1 previous terms, then changes the factors so they are those for the $(n+1)$st term and then does this all over again. The programmer thus needs only to write out a single loop of this sequence—the factor multiplication, the addition, and the factor modification—to start the computer on the first term and let the machine itself go round and round the loop as many times as are needed.

To enable the machine to stop the sequence at the appropriate point, and to give it other flexibility, the conditonal instruction is needed. In essence it tells the machine to take one or the other of two alternate program paths, depending on whether the result of a certain subtraction is positive or nonpositive (zero or negative). This simple property, the consequence of von Neumann's suggestion, gives the computer a degree of logical capability. It can, for instance, stop the summation when an individual term in the series becomes smaller than a given value. Or it can sort out a series of numbers, or of words (for a word to the machine is just another number) and reorder them as desired. As we will indicate later, the conditonal instruction endows the computer with a limited degree of adaptability and freedom of choice.

Of course, with all this seeming freedom, the machine is just following the program written for it. In accordance with the particular sequence of code numbers it has in its memory, it will carry out sequential loops of operations, change to other loops, carry out loops of loops, and finally cause the answer to be printed out. The sequence of code numbers, called the program in machine language, is a long and carefully ordered list of instructions, specifying minutely every possible action the computer is to take. In the early days of computer development, a mere 15 years ago, these programs had to be worked out by some programmer, in complete detail for each problem to be solved. The programmer had to learn a new language, called machine language, in which to write these instructions, so that the machine could understand. It was a tedious job, and one which required a great deal of training to write out such a program without making some slip in bookkeeping or in logic. Complex programs usually needed months of testing and correcting before they worked.

We now realize that such mental drudge work is not necessary, for the computer itself has the capability of writing its own detailed program, of carrying out the bookkeeping and writing the thousands of instructions in correct order, in correct code, once it is told what general operation is required. In loose terms, the machine is capable of translating the user's instructions from a vocabulary easy for the user to learn, into the detailed list of instructions in machine language. In order to do this, of course, there must be written for it a program which will translate the general specifications for solving a problem written in the user's language into the detailed program in machine language, which will enable the machine to carry out the solution. Such a translation program is called a compiler. It very considerably eases the problem of learning how to use a computer. Nowadays it is no harder than learning to drive an automobile. Of course the compiler itself is an exceedingly complex program, requiring many man-years' effort on the part of skilled programmers to prepare. Professional programmers now are needed to prepare compilers and other upper echelon programs.

Employing such a person to write working programs, to solve your individual problems, is like hiring an experienced mechanic to be your chauffeur; usually it's better to do your own driving.

We now have a hierarchy of programs for machine control, which can be best illustrated by describing what goes on at a typical university computing center. An individual user, student or faculty member, with a problem to be solved attends a few lectures on programming, or spends a few hours with a manual learning the simplified compiler language and conventions. He then writes out his description of the processes the machine must follow to solve his problem, in much the same way he would have instructed an expert with a desk calculator in the old days. These instructions are then converted to a deck of punched cards which he hands in for check-out.

A large computer operates too speedily for it to be efficient for each problem to be run separately; the time required to read in the cards and print out the answer would be a hundred or a thousand times longer than it would take the machine to do the problem. For machine efficiency, therefore, decks of cards of several dozen users are transferred automatically to magnetic tape before coming into contact with the computer. When this tape is connected to the machine, the computer is at the same time put under the general control of an operating program, which tells it what to do with each of the dozens of sets of instructions from the dozens of users the tape contains. Perhaps the first set of instructions is from a user just starting his problem; it may be a request that his newly written program, in user language, be checked out. The operator program calls in a check-out program which may find a logical error in the sequence of instructions the beginner has written, or some instruction the machine doesn't understand. These errors are recorded in user language on an output tape, to be transcribed later and handed to the user along with his original deck.

The next set may have a checked-out program in

user language, the request being to translate it into a working program in machine language. The operator program then calls in the requested compiler program, which carries out the translation and records the working program on the output tape, to be converted later into a deck of punched cards to be used later in obtaining solutions. Another user's portion of the input tape may be a working program, with the request that a certain number of computations be carried out with it. In this case the working program is read into the highspeed memory and the requested computations are done, the answers being deposited on the output tape.

After all the requests on the input tape are run off and the answers recorded on the output tape, this output tape is removed from the computer, to be converted to typed answers or decks of punched cards, and another input tape, with new requests, is connected. If the off-line reader and card punch are not busy on other jobs the user may have his answer in less than an hour; if the schedule is crowded he may have to wait till next day.

This example of the various types of programs now in use has been needed in order to make understandable three general comments about present-day digital computers. The first is that the soft-ware, as the hierarchy of programs, operating systems, compilers, and so on is called, is at least as important as the equipment itself, the hardware. A shiny new computer, no matter how speedy, without its soft-ware, is about as useful as an automobile without steering wheel and transmission.

The second comment is that the development of the upper-level programs, the compilers, and the more elaborate operational programs require intellectual ability and originality of the highest order. Indeed, the logical questions connected with the writing of a compiler, a program to write programs—may well expand into a new branch of mathematics. At present this sort of programming is an art; we hope it eventually will become a science. The artists, the first-class programmers, are in very short supply;

too many of them are lured away from university research by astronomical industrial salaries, usually to become "computer chauffeurs."

The third comment is that with the digital computer of today we have been forced to favor the computer's efficiency over the user's efficiency. If the user simply wishes to have another set of calculations done on a problem which is familiar to him, he probably will not mind having to wait a day or so to get the results. But if the calculation is a part of his train of thought, and he cannot continue the work until he has the answer, he will wish he had exclusive use of the machine so his answers would come as fast as the machine can provide them—which is quite fast. The trouble is that the machine would then be idle most of the time, for even a computation which would take three solid months on a desk calculator will be completed in a few seconds and, while the user is thinking what to do next and getting ready to ask it to be done, the machine could have been finishing ten or twenty other persons' problems.

This brings us to the third way of describing computing machines, in terms of the way they can be used. Machines are built to relieve us of drudge work. Most machines relieve us of physical drudgery; they extend and improve our strength or muscular dexterity. Computing machines, when we have learned to use them, should perform the same function for our mental faculties. They already can relieve us of some varieties of mental drudgery. In time, they should be able to extend and sharpen our mental creative powers, both in science and in other intellectual fields. For this to be possible, of course, the computer must be at least as responsive to our needs and decisions as a bulldozer or an automobile is to the requirements of its operator. The problems which must be solved before this sort of ready assistance can be achieved are not all related to computer design. Indeed, many of them are logical questions of challenging difficulty; their final solution will require mathematical and logical inventiveness of high order.

The present generation of digital computers has been designed primarily to carry out numerical computation, an activity in which considerable intellectual assistance can be provided without continuous collaboration between the man and the machine. The storage and simple processing of various kinds of information can also be carried out by electronic computers, with little need for continual supervision when the desired information output is simple and standardized. Many varieties of machines are being developed for this purpose. But the computing machines which will work with the user on intellectual tasks, the way many mechanical machines now work with their operators to carry out physical tasks, has yet to be built. All I can report today are the beginnings of attempts at understanding some of the basic logical, as well as design, problems which must be solved before we can build such a computer. Some of these problems have to do with improving communication between the machine and its user; some have to do with determining how far the machine can go in assisting the user's thinking. Most of them are problems of soft-ware, rather than hardware.

An obvious need is a console which will allow easy communication between the machine and the research worker needing assistance. In some cases, the communication can be via typewriter; for example, if the answers are numbers or short statements. In other cases, where the output is a plot, a diagram or a mass of printed material which is to be scanned, the material may best be displayed on a cathode ray scope, with controls to change its size and to move parts on and off the scope. A light-pen is needed for the worker to point to some portion of the scope display or to draw figures or characters on the scope surface and have the action recorded by the computer. These and other useful instrumentalities are available. What is needed first is a better understanding of the computer's capabilities and then a number of bright ideas on how to put these various instrumentalities together so as to achieve the capabilities.

There is also the practical problem of arranging so that many users can collaborate with one computer at the same time. The speed of the larger, present-day machines is such that a single research worker, using the computer in this direct-contract mode of operation, leaves the central processor idle much of the time. The user may wish the numerical answer to a trial calculation, or the series expansion of some function about a new origin, or the contour plot of the function in some region, or a bibliography of articles which might relate to the subject of immediate interest, or the performance of any other mental drudge work related to his immediate chain of reasoning before he loses his train of thought. But once he has the answer, he may wish to think for several minutes before he next calls on the machine.

Thus, the central portion of the computer, with its processor and hierarchy of memory banks, can serve a number of consoles in parallel, without appreciable interference. The Computation Center at MIT has been operating such a time-shared system most of this year on an experimental basis, using an IBM 7094 as the processor with a disk memory to store one user's data and programs while the computer answers another user's questions. Even with the present rather slow rate of transfer from disk to core memory, as the machine switches from one user to another, we find that between ten and fifteen consoles can be sharing the machine at one time without any single console user being subject to more than a few seconds' delay in machine response. Perhaps you can realize that the operating program for a time-shared system, which must keep 20 users' programs strictly separated and must know when to cut off one user and which user to go to next, is truly a top-echelon program.

The system in operation at MIT[1] for the past ten months uses ordinary teletype consoles which can be placed

[1] F. J. Corbato et al., The Compatible Time-sharing System (Cambridge: The MIT Press, 1963).

anywhere a telephone line can be run. Some 30 of them are in various offices around the Institute; an experimental one is located upriver at the Harvard computing center. Temporary connections have been made for demonstration purposes as far away as Chicago and, just recently, by telephone cable to Oslo, Norway.

In this system the user, when he wants another computer run, can turn to the console in his office, activate it, and type in the code number of his program. The machine types back WAIT, while it gets his program from the disk file—a matter of a second or so—then types READY. He then types in the requirements for the next computation, with parameter values, in a code devised when the program was written, and the machine answers WAIT. If the calculation takes ten seconds (the equivalent of three desk-calculator-months) he may have to wait 20 seconds or so before his answer comes back, if others are using the machine just then. But it usually seems to him that he has the exclusive use of the computer. He can, if he wishes, write and correct his program on the console, or he can turn it in, in card-deck form, to be placed in his part of the disk file.

We find this immediate response, in contrast to getting the answers next day (as so often happens in usual batch-process computation), produces big changes in the way the machine is used and may further improve machine capacity. In the first place the user tends to shorten his requests so he doesn't have to wait long for an answer. In practice the majority of computations requested by time-sharers take less than 15 seconds of computer time. If the computation takes more than a minute, the user is usually willing to let it be fitted in as "background computation," to be done in the odd pieces of time when no one asks for service, and to be finished in an hour or so. The time-sharer has already come to value the speed of response; he gets impatient if he gets no answer in a few seconds. When the response comes, he can think for as long as he wants before he makes his next request without feeling that he is wasting machine time, for meantime the machine is answering other requests.

What kind of mental assistance can we expect from a time-shared computer? How far can the computer go in helping us with our mental problems? We don't know the answer to these questions. I submit that the research needed before they are answered will be both difficult and exciting. The work already started impinges on many aspects of mental activity; it promises to help us learn more about how our own minds work in addition to providing us with an assistant of incredible speed and accuracy.[2] Let me touch on a few recent developments.

At present, users' instructions are fed to the computer in the form of punched cards or tape or, in a time-shared system, by typing on a typewriter directly connected to the machine. One can think of many other desirable ways of communication; nearly all of them are being investigated, some with success. It would, of course, be useful for the machine to read directly a printed or typed sheet of paper rather than having to transcribe the instructions on a card punch or a console typewriter. The technical difficulties involved in reading any specified type front are not great, but it is not certain how valuable such limited capability would turn out to be. It would be much more useful for the machine to be able to learn to recognize whatever characters or figures happen to be involved in a set of instructions. For example, many machine programs can be most efficiently represented in the form of flow diagrams graphically portraying the logical structure of the operation to be performed. In these cases it would be convenient to be able to draw the flow diagram either by light-pen on a cathode ray scope, or by any other graphical means of communication with the machine, labelling the blocks by number or letter and adding instructions by

[2] Automata Studies, ed. by C. E. Shannon and J. McCarthy (Princeton: Princeton University Press, 1956), is a survey of early work in the field. Also, Marvin Minsky, "Steps Toward Artificial Intelligence," Proc. I. R. E., 49, (January 1961), p. 8. This is a review of recent work and includes an exhaustive bibliography.

auxiliary typewriter. For this to work the machine must be able to read hand-printed letters.

The basic problem, that of character recognition, is under active investigation at present. It involves questions of topology as well as those of efficient coding. The letter or geometric figure must be represented in the machine's memory as a sequence of binary digits and the machine must be "instructed" to correlate this sequence with some number or letter it already "knows." One way to encode a character is to break the area it occupies into subareas, each corresponding to a location in machine memory. If the character enters the subarea, a 1 is placed in that memory location; if it does not, the location has a zero. If the grid is too fine, too many memory locations are required for each character; if the grid is too coarse, different characters will not be differentiated. Moreover, unless some standardization of size and orientation of the drawn character is adopted, each character will be represented by a whole series of different code numbers, depending on its size and location in the field of view. In many projected solutions of this problem, the degree of detail required to differentiate between characters is left to the machine itself to determine experimentally. To put it in more picturesque language, the machine is "taught" to distinguish between the letters.

The program recently written by Teitleman[3] at MIT is a promising example of work in this field. His program enables a user to teach the computer to recognize upward of 50 different characters, written by light-pen on any designated area of a cathode ray screen. It takes advantage of the sequence of actions involved in writing a letter. For example, in drawing a capital K, one may habitually start by drawing the vertical line from top to bottom; one may then lift the pen and afterward draw the two diagonal lines in one continuous motion or, alternatively, with one further lifting of the pen.

[3] Warren Teitleman, "New Methods for Real-Time Recognition of Hand-drawn Characters", (Master's Thesis, MIT, June 1963).

In the program of Teitleman, the designated area is divided into four or five different designated subareas, each with its complement, such as upper half (with complement lower half), left third (with complement right two-thirds), upper right corner (complement the rest of the area), and so on. For each of the designated subareas the machine records, sequentially as the letter is drawn, whether the pen is drawing in the area (call this i, for the moment) or is drawing outside the area, in its complement (call this o) or whether the pen is lifted and being moved to start a new line (call this w). For example, the sequence for capital K, for the upper half as the designated area, may be iowio, a new symbol being added only when the pen enters or leaves the subarea or is lifted. Since, in this sequence, no symbol can follow itself, the combination may be recorded by the machine as a binary number; by our rules the initial symbol can either be i, which is recorded as a 1, or an o, which becomes a 0; for the subsequent symbols, following a w, record an i as a 1, an o as a 0; following an o, record an i as a 1, a w as a 0; following an i record an o as a 1, a w as a 0. Thus the sequence iowio becomes the number 11011. For each letter drawn and for each subarea chosen, the machine thus records a binary number.

The procedure of "teaching" the machine to recognize the character is carried out as follows: The letter capital K, for example, is drawn in the designated area on the scope and the machine computes a binary number for each of the chosen subareas. The letter capital K is then struck on the typewriter, thus telling the machine that the set of binary numbers it has computed is equivalent to the symbol "cap K" it receives from the typewriter input. Each letter in the desired set is similarly drawn and identified. By this time the machine has amassed, in its memory, for each different binary number and for each different designated subarea, a list of characters to which it may correspond. For example, for the upper half as the subarea, connected with number 11011, is the list K or B or R, but the list would not include I or E; for the left half as subarea, the list corresponding to number 10011 would include K but not B or R. If the subareas have been chosen correctly the set of numbers will have lists which have only one character in common, in this case K.

In principle, one could work out the minimal number of subareas and their shape, which will uniquely distinguish between any given set of characters. But this is a piece of mental drudgery which we may prefer to let the machine help us carry out. So we choose a set of subareas which look as though they would work, and call on the machine to help test whether this is the case. For the chosen subareas we go through the list of characters and identify them all as just described. Next, we go back, draw one of the characters, and ask the machine to identify it. If we have drawn the character the same way the second time and have been lucky in our choice of subareas, the machine will respond successfully. But we may have drawn the letter somewhat differently the second time or else there may be an ambiguity in the system; 2 and Z may be represented by the same numbers for all subareas. In the first case, there is no character in common in the resulting set of lists; in the second case, two characters are present in all the lists. In this second case, the characters may be distinguished by adding another subarea, sensitive to the difference; to distinguish between a 2 and a Z, for example, we could use the upper, right-hand corner of the space.

In the first case, where one subarea is sensitive to variations in the way a letter is written, a teaching reinforcement process can be followed. For example, the second time the letter K is drawn, the machine may report that the lists for the numbers resulting from the subarea α, β, and γ have K in common, but the list for the number computed for subarea δ does not contain K. The machine is then told to add K to the list for δ, which did not originally contain it, and to increase the "weight" of K in the lists for α, β, and γ. If this procedure is repeated several times, many of the variants of inscription are discovered and subareas particularly sensitive to variation for this letter are downgraded in weight compared to the rest. At the end of the "training period" the machine is told, in cases of doubt, to take the character with greatest statistical weight in the set of lists it generates.

The program written by Teitleman along these lines, using four or five subareas, has been quite successful for sets of 40 characters or fewer, written by individual users. Usually a single writing of each of the English capital letters sufficed to produce complete discrimination, as long as the user was reasonably careful in his draftsmanship. Less careful lettering usually required a teaching period of ten minutes or less to get the machine to "read" the list. A ten-minute training period for the Russian alphabet produced a set of machine instructions which yield an occasional error in one character. A similar test, using a Hebrew alphabet, produced a letter-perfect set of instructions. Each of the generated instruction sets, however, were sensitive to the idiosyncrasies of the individual user; one which was letter-perfect for the user who created it would often produce a number of errors if used by someone else. This limitation is not serious for many possible users; each user can have his own "handwriting" stored along with his various other programs, ready to be called out when needed.

Teitleman's program has been quoted in some detail, not because it is the only successful method of character recognition, but because it illustrates a number of general characteristics of the whole field of research I am discussing. In the first place, the machine has taken over rather more of the mental drudge work than one might have originally expected. Instead of working out appropriate subareas ahead of time, we let the machine work out its own recognition program, after which it can help us find out which set of subareas is most efficient.

In the second place, we see that Teitleman's program is, in modern mathematical parlance, a metaprogram. It does not instruct the machine how to recognize any set of characters; it enables the user, working with the machine, to generate his own set of operating instructions, so that the machine will thenceforth recognize the user's chosen set of characters, written the way the user writes them. It is a program which enables the machine to write its own set of operating instructions, in its own language, with a minimal amount of guidance by the user.

In the third place, we see that the program enables the user to proceed inductively, rather than deductively. A few minutes of trial, rather than a lengthy series of mental visualizations, will suffice to show whether a given set of subareas will be satisfactory. And the introduction of the "training" routine, with its procedure of reinforcing successful results, reducing weight for unsuccessful choices, protects the instructions from undue sensitivity to irregularities in operator manipulation.

A less detailed discussion of some other developments in pattern recognition will serve to illustrate other ways in which the computer can carry out tasks of mental drudgery. The high-energy particle accelerators, which nuclear physicists use in exploring the properties of elementary particles, produce large quantities of data, usually in the form of stereo-pictures of bubble or spark chamber tracks of these particles and their reactions.[4] The machines now operating in this country can produce more than 20 million such pictures a year, each of which must be examined to find whether any reaction of interest has been recorded. At present the pictures are individually culled over by technicians who pick out possibly interesting ones and these are analyzed stereoscopically to obtain the charge, momentum, and energy of each initial and final constituent in a reaction. Here, obviously, it a task which could and should be taken over either entirely or in part by a computer, provided someone can program the machine to do the job more quickly and more accurately than the technician.

[4] W. F. Miller and H. W. Fulbright, "Review of Data Analysis Systems for Nuclear and High Energy Physics in AEC Laboratories," AEC Div. of Res. Report (November 8, 1962).

Several projects are now in progress with this goal in view, exploring different concepts of machine use. The basic problem is to relate the confused pattern of bubbles, for example, shown in the usual picture, with the underlying pattern of curved lines, line ends, and line intersections in space which represent the particle tracks and nuclear reactions. It is easy to program the machine to place in its memory the record of a TV scan of a picture. What is desired is a program instructing the machine to convert this mass of data, mostly superfluous or redundant, into a usable record of positions, curvatures, and densities of track and positions and multiplicities of intersections. The record must be structured logically so that the machine can pick out the pictures of possible interest and can present the corresponding data in a form most amenable for analysis of the reaction by an expert.

It is not certain yet how much of this procedure can be mechanized. The conversion of the TV scan into digits in machine memory which can be redisplayed as lines and intersections or which can be further manipulated to obtain energies and momenta has been accomplished in several ways. There is already no doublt that computers can scan pictures rapidly and can cull out those displaying no "interesting" reactions, leaving a relatively small fraction of "interesting" pictures to be looked at by the human operator. It has also been demonstrated that these pictures can be played back on a TV scope, by the machine, in "clean-up" form, with "uninteresting" tracks eliminated and "interesting" ones marked, if need be, for density and curvature, so they can be quickly analyzed by an expert. The presently developed programs will require the full time of one large computer for each of the existing large particle accelerators, but one hopes that clever programming will be able to reduce the machine time required per picture by one or more factors of ten, so that the machine can have time to do other useful work such as helping to analyze the reactions it has recorded.

It is not certain yet, however, whether one can program

the machine to recognize a specified kind of nuclear reaction, or whether it can recognize and call to the attention of the operator a reaction which neither it nor the operator has ever seen before. Such questions probably will best be answered inductively, experimentally, once the meta-programs for track analysis have been improved in speed and efficiency. What the machine then could do would be to search through its data on a set of pictures, which it has already scanned and recorded in manipulable form, to see how many of them represent a particular kind of reaction described by the operator in terms the computer can understand. Whether it can go further than this probably depends on the cleverness of some nuclear physicist who is also a programming expert, to combine physics and machine logic into one neat program for reaction-pattern recognition.

Here again the ultimate range of utility of the digital computer, as a direct assistant in research, depends on its ability to search among vast multitudes of data, to find a number or a pattern or a relationship which satisfies some not-always-clearly-stated criterion. The task we ask of the machine is of the same nature as the task confronting any human attempting to find the solution to some intellectual problem.

How far in this direction can we get the computer to help us ? What sorts of intellectual problems can we expect the machine to solve without our detailed guidance ? This is the basic question. A number of groups are working on various aspects of this question; however, the number is small considering the difficulties involved and the potential importance of the answers when found.

One direction of attack is to study the capabilities of computing machines to play various kinds of intellectual games, such as checkers or chess. The advantage here is that degree of success in finding a solution is easier to measure for a game than for many other sorts of problems. The investigation can thus concentrate on programming the search without having simultaneously to devise complex criteria of successful search. Here one must instruct the machine so it can find, among a large number of possible moves, the particular sequence which will lead to success in the face of the opponent's moves. A program to play as simple a game as tick-tack-toe is not the point here; all possible moves and countermoves for such a game can be stored in memory and the program can be completely determinate, with no requirements that the computer search for a better or a best move.

For checkers, and even more for chess, the number of possible moves is far greater than can be stored in any machine now, or soon to be, available. One must therefore produce and program an algorithm which will enable the computer to start from the present state of the board, at any time during the game, to look ahead two, three, or more moves, and work out the next move which will most surely achieve future success in the face of the opponent's moves. The machine must explore the branches and twigs of a tree, representing all moves possible in the near future, and pick from these the branch which has greatest promise of success and least danger of falling into a trap set by the opponent.

Here again, the more successful programs include routines whereby the machine can "learn" by playing the game, can reinforce those choices which turn out to be more successful, and can call on this past experience in cases where the issue is in doubt. Samuel, of IBM, has written a checker-playing program for a 7090 computer,[5] involving these various features, which has defeated a former state checker champion in a rather well-played game. As one would expect, the extant chess-playing programs are not as successful; chess is a more complicated game and present programs just hold their own with beginners at the game. A program now being written at MIT will, it is hoped, hold its own with players of moderate ability.[6]

[5] A. L. Samuel, "Some Studies in Machine Learning, Using the Game of Checkers," IBM Journal of Res. and Dev., 3, (July 1959), p. 211.

[6] B. H. Bloom, "Proposal to Investigate the Application of a Heuristic Theory of Tree-searching to a Chess-playing program," Artificial Intelligence Project—RLE and MIT Computation Center Memo No. 47 (February 15, 1963).

It will utilize more sophisticated techniques of evaluating the relative strengths of each player at each point in the game (called Point Count[7] Chess) to cut down the amount of blind searching of all possible moves which has made previous programs so unwieldy. It also will include provision for gaining experience by playing games.

Another area in which the logical capabilities of a computing machine can be tested is in the devising of proofs of mathematical theorems. This also involves a search and evaluation procedure; the proof must be built up from axioms and a set of already-proved theorems. One must determine which of these are relevant to the new theorem, how they may be structured to achieve its proof, and, if steps are missing, whether these can be built out of the existing material, and so on. Newell, Shaw, and Simon[8] wrote a program which was able to "prove" elementary theorems in propositional calculus and which has given rise to a lot of discussion and some further development, but the field may be too difficult to expect much progress at present.

Of more immediate practical interest, though perhaps of less ultimate conceptual value, are programs instructing the machine to perform certain mathematical manipulations, such as the expansion of a function, or combination of functions in a power series, or its differentiation, or its integration. A program bordering on language translation[9] is one which instructs the machine how to go from a set of statements in English, such as "John bought five oranges; one orange costs seven cents; how much did John pay?", into a set of algebraic equations, which can then be solved. It is well known that the mental task of translation from sentences to equations is more difficult for the beginning mathematics student than is the solving of the resulting

[7] I. A. Horowitz and G. Mott-Smith, Point Count Chess (New York: Simon and Schuster, Inc., 1960).

[8] A. Newell, J.C. Shaw and H. A. Simon, Emprical Exploration of the Logic Theory Machine, Proc. W.J.C.C., (1957), p. 218.

[9] D.G. Bobrow, "A Question-answer for Algebra Word Problems," Artificial Intelligence Project—RLE and MIT Computation Center Memo No. 45 (January 1963).

equations. The translation from verbal to mathematical instructions is certainly a part of the process of easy communication between user and machine; its possible difficulties should be explored.

Language translation itself is being intensively investigated; since a compiler is a program to translate from user language to machine language, it is not suprising that some enthusiasts have started work on translation programs for the immensely more intricate and less logical "natural languages." The subject is too large to discuss here beyond the comment that even if the actual translations may not ever be smooth or polished the research will undoubtedly teach us more about the structure of natural languages than we would have learned otherwise.

This whole incompletely defined field of exploration of machine capabilities in various aspects of intellectual activity has been given the rather pretentious name of Artificial Intelligence. The number of research workers engaged in its exploration are few as yet, but the challenge of the problems it contains and the obvious potentialities of the results which might emerge are attracting an increasing number of bright graduate students. One can confidently expect that this next generation, being as familiar with machine programming as with classical mathematical analysis and logic, will provide the symbolism and the algorithms which we now can see are needed to make electronic computers the intellectual counterparts of our ubiguitous mechanical assistants.

Already we can see some of the possibilities when we have somewhat larger machines that can accomodate several hundred time-shared consoles. Some of them could be in libraries. At MIT we have a program which will on request type out the title and location of any article written by a given author and published in, say, the "Physical Review" in a specified year or years. Or the user can type in the title and author or author and location of an article and request the titles, authors, and locations of all articles which referred to the article typed in. Or he could request a list of those articles which refer to at least two, say, of the references in the article typed in (this last is a quite efficient way of finding articles on closely related subjects). The library hopes later to record its book circulation and to type out its overdue notices by similar or specialized consoles. Many other library reference programs are being discussed or investigated.

The possibilities in education are also as yet practically untouched. But enough preliminary work has been done to show that within a few years a student can go to a console and request a drill-routine in calculus or in irregular French verbs. The machine will respond with a sequence of questions, which the student answers. Depending on the correctness of the answers the machine will progress fast or slow, will hit the high spots or will repeat with variations until the answers are all correct. If desired, the student's score can be recorded and averaged or graphed for analysis by the instructor.

But the possible applications in industry are just as exciting as those in education and research. The airline reservation service now in use is a very specialized, time-shared system. One could imagine a centrally located computer being fed a company's production, inventory, and sales figures each day from many consoles spread over the country. It could, almost simultaneously, tell some warehouse manager where the nearest supply of part X is, or tell the production manager what would happen to the production schedule of product Y if factory B were to concentrate on some new product for a month. And the sales manager could display last week's (or yesterday's) sales figures on a map of the United States during his report to the executive committee. Or he could request the machine to work out the cost and possible results of a suggested advertising campaign on the West Coast, using the latest decision-theory model for the calculation, and get the answer while the committee is still debating the problem.

All of these uses, and many more, will be possible when

we learn to make computers as responsive and adaptable assistants as bulldozers and automobiles are now. The hardest part of the development will not be the design of new equipment, but will be the mental effort of developing new programs and system-designs to improve machine response, and the intellectual dislocation required to visualize new tasks for our new assistant.

The Revolution in Biology and Medicine

Bentley Glass

Bentley Glass, *born in China, received his Ph.D. in genetics from Texas in 1932, and afterwards was a National Research fellow in genetics at the University of Oslo, Kaiser Wilhelm Institute, and the University of Missouri. His teaching career started at Stephens College in 1934; he has taught since at Goucher and at Johns Hopkins where he remained until 1964 when he became Academic Vice President of the State University of New York at Stony Brook. He is editor of the* Quarterly Review of Biology, *biological editor for Houghton Mifflin, and a member of the editorial board of various biological journals and compendia. He has been President of the American Institute of Biological Sciences, and Chairman of the Biological Sciences Curriculum Study. He served as a member of the Advisory Committee on Biology and Medicine of the Atomic Energy Commission, and of the Committee on the Genetic Effects of Radiation of the National Academy of Sciences. He is a member of the National Academy of Sciences, the American Philosophical Society, and the American Academy of Arts and Sciences. He is a Senator and Vice President of Phi Beta Kappa, past President of the American Society of Naturalists and the American Association of University Professors, President-Elect of the American Society of Human Genetics, and a member of the Board of Directors of the American Association for the Advancement of Science.*

The Revolution in Biology and Medicine

THIS IS SURELY the century of science in human history. I often wonder whether we are quite aware of what times we are living through. Over the past decade I have become increasingly concerned about the importance of the sciences in liberal education. One of the things that impresses itself upon me more deeply with every passing year is the real significance of the sciences in history. Indeed, I wonder whether, in the next century, this century of ours will be known as the century of the two great World Wars, the century of Hitler and Roosevelt, Churchill and Stalin, Kennedy and Khrushchev, or whether it will not perhaps more likely be remembered as the century of Einstein, of Muller, and of the advent of nuclear physics and molecular biology. That may be a very bold claim, yet as we look back upon the nineteenth century do we really think of it as the century primarily of Queen Victoria and Gladstone, even of Lincoln and the Civil War, so much as we think of it as the century of Louis Pasteur and Charles Darwin? Yet far more than the nineteenth century, the twentieth century is preeminently the century in which science is the principal current of human history, the principal factor producing changes in the conditions of human life, the principal factor that must be reckoned with by all of those who teach in universities, whether they teach history or philosophy, languages or literature.

During the past seven years—that is, since the beginning of 1959—I have been responsible for the policies and guidance of the Biological Sciences Curriculum Study, a program which was undertaken in order to bring up to date and to improve in scientific outlook the teaching of biology in the secondary schools of our country. The immediate stimulation of the governmental support for that program developed from a growing recognition, after World War II, that the military security of the nation, as well as its rate of advance in the technology of modern civilization, depends directly upon scientific discoveries and that, in turn, the rate of scientific advance depends upon the level of education in the sciences. Nevertheless, there has long been emerging in our general community a realization that something radical must be done to cope with the ever-widening disparity between education in the sciences and their actual status. In biology the textbooks currently studied in our schools represent for the most part the biological knowledge of 30 years ago. Thirty years ago is not actually so very far in the past. It was the time when I began teaching as a college professor. It was the time of the Great Depression just before World War II. And yet what has happened in the sciences, and particularly in the biological sciences, during the past 30 years has made what was taught as a summation of the sciences 30 years ago seem actually as antiquated as the science of the mid-nineteenth century.

To see this problem in perspective, one might note that according to one recent estimate, the number of scientific publications is doubling every 12 to 15 years. If our fund of biological knowledge, measured by these publications of new research and discovery, was four times as great in 1930 as in 1900, then in 1960 it is 16 times what it was in 1900 and in the year 2000 it is likely that it will be 100 times as great. One may feel rather sorry for the students in the year 2000 who will need to learn 100 times as much as we could have expected students to learn about the sciences in the year 1900. But of course it cannot be done. There must be a selection, a synthesis, of this scientific knowledge, and not merely an effort to teach all that was known in the past plus all that has been discovered in each succeeding decade. To cite another startling estimate, we might note that 80% of all the scientists who have ever lived are living and producing now. At the present rate of increase in their number, the million scientists of today may well increase to 20 or 30 million by the end of the present century.

Another, perhaps more vivid, way to look at the problem is to consider what might have been found in a good textbook of 1900 compared with one of 1930 and one of today. In biology, for example, in spite of the very great achievements of the nineteenth century—the cell theory and nature of cell division, the cellular basis of reproduction, evolution theory, the germ theory of disease, and the foundations of modern plant and animal physiology—the textbook of 1900 would be scanty indeed in its coverage of areas of major importance. There would be nothing in it about genetics, for Mendelian heredity was not rediscovered until the year 1900. Biochemistry in its modern sense had begun only three years before, far too short a time for it to get into the textbooks. Eduard Buchner's classic studies on the nature of the enzymes had just begun. The first hormone was not discovered until after that year. Immunology did not exist. It was in 1900 that Karl Landsteiner discovered the blood groups, and the existence of antibodies was first made known by Emil von Behring and S. Kitasato only ten years before. It is very unlikely that the book would have hinted that there were such things as viruses, for although D. Iwanowsky had discovered the first plant virus in 1892, animal viruses — which cause so many of our own diseases — were unknown until 1898. Nothing was known about specific vitamins in 1900. The experimental study of the factors governing the development of the embryo was just beginning. Pavlov had yet to do his classic studies in the fields of physiology and experimental psychology, the studies of conditioned reflex behavior in the dog.

TWENTIETH CENTURY ACHIEVEMENTS

Thirty years later all of these studies were well advanced, and a large number of new biological sciences had sprung into being. The postulation of the chromosome theory of heredity by W.S. Sutton and Theodor Boveri, and its subsequent proof through genetic work with *Drosophila melanogaster* by Thomas Hunt Morgan, A.H. Sturtevant, H.J. Muller, and Calvin B. Bridges constitutes one of the towering achievements of science in the twentieth century. Almost equally significant was the unraveling of multifactorial inheritance and the application of this to hybrid vigor by E.M. East and G.H. Shull. This development led directly to the production of hybrid corn, a truly magnificent contribution to agricultural productivity.

Another area of outstanding achievement in these first 30 years of the present century lay in the field of nutrition, where early indications that food substances besides protein, fat, and carbohydrate were required for a complete, health-promoting diet led to the discovery of the vitamins. The investigations of F.G. Hopkins, C. Funk, and E.V. McCollum were but the most outstanding of a host of researches that elucidated the nature of vitamin A, thiamin, riboflavin, vitamin C, vitamin D, vitamin E, and nicotinic acid, and ascertained the deficiency diseases each of them corrected.

In the study of disease and resistance to disease great progress was made. Animal viruses were discovered and in 1918 d'Herelle added the bacteriophages, or bacterial viruses. Antigen–antibody relations were intensively explored and a whole new field of immunology developed.

Unexpected discoveries were also made. In 1927, Hermann J. Muller and Louis J. Stadler independently discovered that high-energy radiations such as X-rays will produce mutations in the genes of animals and plants, and thereby placed in the hands of man a really formidable power to alter the agents of heredity. In that same year G.D. Karpechenko achieved the creation of the first artificial new species created by the hands of man, by utilizing a method which nature itself has employed in the origination of many of our most important cultivated plant species—cottons, wheats, tobacco, and others. In these same years R.A. Fisher, J.B.S. Haldane, and Sewall Wright laid the theoretical foundations of a new development of evolutionary theory, based on genetics, susceptible of experimental testing, and adding to the methods whereby man can modify genes, select new types, and create new forms of life on the earth—and giving him also the power to meddle, wisely or unwisely, even with his own hereditary nature.

Also during the 1920's, Hans Spemann's experiments were disclosing some of the remarkable elements in the control of the development of the animal embryo, especially the action of the organizer responsible for the induction of the nervous system in its characteristic location. Similarly, new insight into the factors controlling plant growth came with the discovery in those years by P. Boysen-Jensen and Frits Went of the growth-accelerating substances in plants, the auxins.

THE EXPONENTIAL PROGRESS OF BIOLOGY

Now consider the past three decades of progress in biology. In spite of the interference with biological research caused by World War II, the advance has been well-nigh incredible. During these years we have learned how to induce mutations with a great variety of chemical agents through pioneer work by J.A. Rapoport in the USSR, Charlotte Auerbach in Scotland, and F. Oehlkers in Germany. Especially significant is the indication that certain agents, unlike ionizing radiations, may be used to produce only characteristic types of mutations, and hence that by means of the study of directed mutations we may analyze further the structure of the hereditary material itself.

It was during this period that the study of the transformations of *Pneumococcus* cells by material extracted from cells of a different genetic strain proved that the hereditary material in bacteria is deoxyribonucleic acid

(DNA). Another classic experiment demonstrated that for the phage infecting *E. coli* only DNA is injected into the host bacterium, while virtually all protein remains outside. This evidence radically shifted the previous opinion that regarded protein as the replicating material basic to the nature of heredity and reproduction.

CRACKING THE GENETIC CODE

In the fifties came the fruitful theory of J.D. Watson and F.H.C. Crick regarding the structure of the DNA molecule and its method of replication. The famous hypothesis of the double helix with its paired purine and pyrimidine bases has led to much vigorous investigation devoted to the problem of how DNA passes information to RNA (ribonucleic acid), and how the latter in turn specifies the amino acid structure of protein or polypeptide chains.

Arthur Kornberg has shown that it is possible to take from the common human intestinal bacillus *Escherichia coli* an enzyme that, when provided in a test tube with four fundamental kinds of nucleotide triphosphates and a bit of native DNA extracted from some living cell, will synthesize an abundance of new DNA, of the same kind as the primer or model DNA put into the system. This is the first essential step in the artificial creation of genes by man. Marshall Nirenberg and Severo Ochoa, working independently, have in 1961 "cracked the genetic code" and carried to completion a scientific investigation analogous to the deciphering of Egyptian hieroglyphics, Babylonian cuneiform, or Minoan B. That is to say, they have determined the composition of the 21 nucleotide triplets in ribonucleic acid which convey the genetic information from the DNA to the protein-synthesizing centers of the cell, and there specify which of the 21 usual amino acids is to be inserted, in its proper position, in the protein molecule being synthesized. Frederick Sanger has used a most amazing type of analysis to dissect, by means of proteolytic enzymes, the entire protein molecule of insulin, to identify its 51 amino acids, and to map the entire sequence of these within the molecule. Since that work, several more protein molecules have already been completely finger-printed in a similar way.

The number of vitamins formerly known has rapidly increased during these 30 years—folic acid, vitamin B-12, pyridoxine, nicotinamide, and others—and, what is of greater importance, we have learned something more significant about them than that they are required in human diet. We have learned that vitamins are commonly incorporated into coenzymes, each of which plays an essential, very specific, role in the chemical machinery of life. It was during the 1940's that Fritz Lipmann first pointed out the remarkable nature and functions of adenosine triphosphate (ATP). Today we can scarcely conceive how it was ever possible to talk about human metabolism, or the use of energy in any plant or animal or microbe, without a knowledge of the role of this substance. Vincent du Vigneaud has shown how to synthesize some of the polypetide hormones produced by the pituitary gland. They were the first hormones actually to be synthesized. A. Butenandt, Edward A. Doisy, E.C. Kendall, and other workers have studied the steroid hormones of the adrenal cortex and gonads and found the way to synthesize them artificially. Out of these investigations came the remarkable studies of the physiological action of cortisone, of ACTH, and of the sex hormones.

During these years, the efficacy of the sulfonamide drugs was discovered. Sulfanilamide had been known for decades as a chemical product, no special use for which had been found. Now it was first tried out as an antibacterial agent and proved to be tremendously effective against many different kinds of bacteria, since it competes metabolically with a certain coenzyme (para-aminobenzoic acid) necessary to the microbial organism.

Antibiotics were discovered in this same period: penicillin, streptomycin, aureomycin, and a host of others. It is already hard to think back to the time when we didn't have this armamentarium of antibiotics able to banish infectious disease almost completely, if we but use them wisely. Nor should we forget the development of the new insecticides, the most widely known of which is DDT, enabling a vastly greater control over typhus fever, malaria, and flyborne diseases to be introduced. The effective control of many plant pests has likewise been greatly aided, and agriculture benefited tremendously.

The introduction of the electron microscope and the phase microscope during these years had led to a resurgence of morphological studies of cell structure and process. We may now hope to tie together the biochemistry and the ultrastructure of the cell. Especially promising in this respect are studies of the chloroplasts, throwing new light on photosynthesis, and of the mitochondria, illuminating the basis of respiration.

We may locate Hans Krebs' citric acid cycle, discovered in the 1930's, in the latter; we may place the steps in CO_2 fixation, worked out by Melvin Calvin, and of photosynthetic phosphorylation, discovered by Daniel Arnon, all in the years of this last decade, in the former. For the first time, too, we can now visualize from photographs the structure of the viruses, and even photograph the single large DNA molecule found to constitute the entire hereditary material of some of the bacteriophages, or bacterial viruses.

In these same 30 years, Wendell Stanley succeeded in extracting enough tobacco mosaic virus from infected tobacco leaves to crystallize it. The studies of the properties and effects of pure virus threw new light on the borderland between living organisms and the inorganic world.

Some of the discoveries made in this most recent 30-year period might seem to be purely of academic interest, for example, the discovery by Joshua Lederberg and Edward L. Tatum that bacteria have sex and will mate if you get the right sorts together. Yet this discovery opened up the entire field of bacterial genetics, previously a mystery. It led to the discovery that bacteria have chromosomes like those of higher organisms, with genes arranged in a linear series, and that these genes govern the biochemical processes of the organism's metabolism in essentially the same way as in human beings. A large part of human biochemistry and pathology we now owe to studies made first on bacteria.

The basic analysis of this sort of study was developed earlier during these three decades in the work of George W. Beadle and E.L. Tatum on the effects of mutations in the pink bread mold, *Neurospora*. As the British physician A.E. Garrod had predicted in 1908 on the basis of certain human "inborn errors of metabolism," as he called them, Beadle and Tatum found in *Neurospora* that each individual mutation that prevented growth did so by blocking a single specific step in metabolism. Through comparison of the metabolism of mold, bacterium, and man, we now stand at the dawn of a new era in medicine.

The study of infectious diseases is no longer so important as once it was. Many physicians in the countries with advanced health standards have never seen a case of malaria, typhoid fever, and other diseases that were once the scourge of mankind. But we still have our inborn errors of metabolism, the faults of our genes, and now we must begin to learn, step by step, how to ameliorate them. This can be done when we learn exactly what metabolic step is blocked in each disorder, what essential product of that step must be supplied, or what substrates, accumulated in the body through lack of normal utilization, are producing harmful effects and must be eliminated.

Biochemistry and genetics have joined forces in a new type of medicine that will clearly have for the future as great a consequence as the discovery a century ago that bacteria and viruses are the agents of infectious disease.

These 30 years saw tremendous strides in the development of evolutionary studies. They advanced from a theoretical and observational basis to many kinds of experimental analysis and verification. A catalyst in this respect was the appearance in 1937 of a book entitled "Genetics and the Origin of Species" by Theodosius Dobzhansky. Equally influential has been A.I. Oparin's "The

Origin of Life on the Earth" (1936). The influence of these works has spread more and more widely through the biological sciences, until the dormant—some persons said moribund—Darwinism of the early 1920's has developed into a strongly knit, experimentally based science permeating all of biology. Its influence is equally clear in paleontology (G.B. Simpson's "Tempo and Mode in Evolution" and later works), in zoology (B. Rensch's "Neuere Probleme der Abstammungslehre"), and in botany (G.L. Stebbins' "Variation and Evolution in Plants"), to name but a few examples. It has given new force and direction to systematics. It has invigorated the study of anthropology and human evolution, and is partly responsible for the great current interest in the blood group systems which led to the discovery by A.C. Allison that the "harmful" gene which produces sickle hemoglobin is maintained in central African populations because it confers protection against tertian malaria, and the equally novel and exciting finding by Vernon Ingram that sickle hemoglobin, the product of a single gene difference from normal hemoglobin, differs from normal hemoglobin in a single amino acid residue in the entire molecule. Thus we learn that the gene acts by specifying the order of the amino acid groups in a protein molecule.

Space and time, and my own knowledge, fail; and a more exhaustive account of even the most outstanding biological developments of these 30 years might lead us too far from the point to be made here: the exponential increase of scientific knowledge in our time. Suffice it to say that in ecology and in the study of animal behavior, in the study of populations and communities and their management, as well as in the responses of whole organisms to their environment, and in every aspect of their biology down to the biochemistry and biophysics of their ultimate atoms and molecules, there have been vast and fundamental advances of which note should be taken.

What the remainder of the century will disclose, we can only guess. We can, however, be confident that our scientific information is likely to increase at the same formidable exponential rate. Along with the practical and theoretical kinds of discoveries just recited, there has been a very great increase in our insight into some of the problems that will face man biologically in the future, especially an understanding of race and of evolutionary problems that we may hope will enable us to act more wisely in the future.

What will biology be in the year 2000? It is rash to guess, for what biologists of 1930 would have dreamed at that date of our present knowledge and perspective? Yet, perhaps, we can see a few directions in which our control over life will be extended. Let us speculate. We will probably learn through increasing studies of the processes of aging how to extend the life span somewhat, but of vastly greater importance, we will discover how to maintain the vigor of mature life into advanced years. Already it is evident that a person of 60 or 70 years of age in our present environment manifests, as a rule, the vigor and interest in continuation of active work that was characteristic a decade or two back only of persons considerably younger. I think we might look forward, before the end of this century, to a time when a person of 80 or 90, or even 100, years of age will still be mentally active and alert, interested in work and contibuting to the intellectual life of society.

I suspect that by the end of the century biologists will have learned how to create some simple forms of living organisms, perhaps at the level of complexity of a virus. To do that is not too far beyond the recent preliminary discoveries of how to synthesize DNA. By the year 2000 it should be feasible to cultivate human reproductive cells, as well as those of animals, in artificial cultures. Some of us have already been trying to do this, often with little success, but the difficulties appear to be merely technical. Sooner or later we shall surmount them. We would then be able to produce normal embryos *in vitro*, as the biologist says, and to raise them to term either in artificial cultures (the "bottle babies" of Aldous Huxley's "Brave New World"), or by implanting them in foster mothers. If, in doing so, we can use the method of transforming bacterial cells by means of DNA extracted from donor cells, it may become possible to modify defective genes or to replace them with sound ones. This possibility has been called "genetic surgery." Even more readily, since genetic studies will have advanced to the point where we can determine the carriers of many of the harmful genes which are present in the human population in a heterozygous, somewhat hidden, condition, it will be quite possible for each of us to go to a heredity clinic before being married and undertaking to rear a family, so as to decide what particular harmful genes we carry and thus avoid producing offspring by unions with individuals who carry the same defective genes we do. This is very worthwhile, since nearly all of us carry several defective genes—four to ten apiece is the best estimate. So far as that is true, none of us is a completely desirable donor to the next generation. Whether or not we can serve society well as the progenitors of future generations of human beings depends largely upon what matings occur. In future decades there must be far more guidance in these respects than there is at the present time. Whether advisory or compulsory, that will depend upon the kinds of social systems we evolve.

Before the next 40 years have passed, we will surely have solved the problems of photosynthesis and have learned how to assure mankind of an inexhaustible supply of food. That food may not be very palatable, at least at first, but biologists of the future, given time, can provide even good flavor as well as sufficient quantity.

Man will certainly have acquired the power to accelerate his own evolution. The crucial question is what direction he will choose, and that question is not one solely for the biologist—in fact, perhaps not for the biologist as such at all—but for all society to choose.

Infectious diseases along with hunger will very likely have been banished completely by the year 2000, but sufficient space for an uncontrolled growth of population seems out of the question on our little planet. One mathematical estimate indicates that a hundred individuals, 50 males and 50 females, reproducing at a 1% rate of increase per annum, a rate only a little more than half of the present rate of increase in the United States, could in 5000 years, a mere fraction of the history of man as a civilized being on the earth, produce 2.9 billion people per square foot of the earth's land surface. At our present rate of population growth, the United States population will in less than a century exceed the present population of China—800 million persons. The population density may be unendurable; we will vacation in crowds and sit for hours in traffic jams and live literally on top of each other. Obviously, something has to "give." Sufficient space for uncontrolled growth of population is out of question on this planet, and happiness is not something that flowers inevitably from the burgeoning of science.

What we most clearly need, it seems to me, is an earnest consideration in this next 40 years of our century, of the relation of scientific advance to the problems of social growth and organization—a real combining of the efforts of scientists, on the one hand, and of social scientists, humanists, and politicians, on the other. We need statesmen of science, and we likewise need statesmen of affairs who are not antipathetic to science but who understand the problems that science is introducing into modern life.

THE TRANSMISSION OF OUR SCIENTIFIC HERITAGE

I turn next to the problem of the transmission to the next generation of the scientific heritage we have produced. Our educational system is very poorly organized to cope with the transmission of a selected core of facts, concepts, and principles derived from an exponentially increasing body of knowledge.

In the first place, consider the teacher. The teacher of today is the student of yesterday who passes on largely the knowledge he learned himself when a student. If the teacher continues in his profession for 30 or 40 years, his skill in transmitting knowledge may greatly increase, but his stock in trade is outdated by a generation or two. Our science textbooks—for example, the newest ones we have just painfully brought up to date through the efforts of the Biological Sciences Curriculum Study and of which we are inordinately proud—will be out of date in five years and urgently in need of revision. If they will be essentially obsolete in five to eight years, how long do you think it takes for a teacher to become obsolete ? And what are we doing about it ? The situation is aggravated because the secondary school teacher learns from a college or university teacher who in his turn, unable fully to keep up with the advancement of science, is also in danger of presenting an antiquated picture of it. One does not escape this dilemma by adverting to the writers of the textbooks, for they are in the same position as the rest of their colleagues. Unless they remain avid learners throughout life and unless they acquire rare skill in critical evaluation and synthesis, they too suffer from a horrifyingly rapid rate of obsolescence—about equal to that of an automobile.

Our curriculum studies in the natural sciences represented the first answer to a growing awareness of this problem. The curriculum studies began in physics and extended to mathematics, biology, chemistry, and earth sciences with the support of the National Science Foundation. For the first time in the history of American education, these curriculum study groups are doing more than merely planning, outlining, and recommending what should be done. They are actively preparing materials for use in the schools—textbooks, laboratory programs, new innovations in teaching—and are engaging as wide and expert a group of scientists and teachers as can be enlisted for the purpose. One of the most encouraging things in my whole experience has been the opportunity to watch how university and college biologists have cooperated with selected high school teachers in the preparation of these books and laboratory programs.

Those are first steps, and of course it will remain necessary—such is the tempo of scientific advance—to revise the new books almost as soon as they have been issued. With the introduction of truly modern teaching materials, a new problem has arisen. The teachers in our schools are not at present trained in the basic sciences they need to adopt such a program and to teach it well. Over and over, we hear that the new books are fine for the students, but the teachers cannot handle them. New training institutes are necessary to reeducate our teachers in their knowledge of present-day science. For example, many of those trained to teach biology do not know enough chemistry, particularly enough biochemistry, to deal with modern biology. Gradually we are awakening to the need for frequent and systematic continuing education in science for all those who teach it.

The summer science institutes begun by the National Science Foundation in the United States have been helpful, but by no means meet the need adequately. Most of the teachers who go to summer science institutes become attracted to this way of spending a profitable summer and return year after year, while the great majority of the teachers, especially those who need help most, never go at all. They cannot even get accepted into the summer science institutes.

I have been greatly encouraged by watching a new movement in this area which has recently developed in Japan. In each of the prefectures of that country, 39 in all, and at local demand, a science education center has been established. All elementary and secondary school science teachers are being required to attend these science education centers for seminars, short courses, or half-year courses on leave from their schools in order to engage in retraining, to learn what the new science is like so that they can go back and teach a science that is not antiquated. How desperately we need a system like this in our own land!

It has already become evident, too, that if the teaching of biology and the other sciences in the secondary schools is to proceed very effectively, there must likewise be a thoroughgoing revision of the program of science teaching in the elementary school and junior high school levels. That too will be undertaken. Can the university remain dormant? I have been very keenly interested in the attitude of university teachers of biology to the new developments taking place in the science curricula of the high schools and elementary schools. Many university teachers are interested in what is happening, but the general impact on the universities up to now has been negligible. Most universities have undertaken very little, if anything, more than to provide better courses for training graduate students to become specialists in the sciences and to add to this volcano of scientific knowledge, the paper of which is burying us just as the inhabitants of Pompeii and Herculaneum were buried by the ashes from Mount Vesuvius. No, we must pay more attention to the teaching of the sciences throughout the curriculum, from the beginning of school experience to the end. And surely, among the most important functions of the university is the development of systems of training and retraining periodically all the teachers of science.

What is the main objective to keep in mind as we undertake this formidable task? In my opinion, it is to learn and to teach the nature of science. James Harlow, the executive director of the Frontiers of Science Foundation of Oklahoma, recently said that probably "fewer than ten per cent of the entire pre-college teaching staff holds any real awareness of the basic nature of science." There are so few science teachers among all teachers, and among the science teachers themselves so few who have acquired a real understanding of the nature of science. And yet, if we are going to develop a civilization broadly and soundly based upon scientific foundations—and we can hardly escape that now—the general citizen of this country, every man in the street, must learn what science truly is, and not just what scientific knowledge can bring about. Surely, this is our primary task. If we fail in this, then within a very brief span of years we may expect either nuclear devastation or world-wide tyranny. For, on the one hand, it is not safe for apes to play with atoms and, on the other hand, neither can men who have relinquished their birthright of scientific knowledge expect to rule themselves. For scientific society to be democratic and to remain democratic, the people themselves must understand the nature of the scientific forces and problems that dominate their lives. For us who teach, this by no means signifies teaching simply a lot of facts about science or even its important concepts and principles—that the earth is round and not flat, that the atom has energy in it, that genes control the paths of development. All those things are important. Yet far more important, it seems to me, is the comprehension by the learner of the true nature of the process whereby men gain reliable knowledge about matter and energy and whereby men acquire new understandings of life and of man's place in the universe. Here lies our power. As teachers and as citizens, we can be pioneers in the effort to reorganize and reinvigorate the teaching of biology and of the other sciences, to give them new direction, and to make of our education for the coming years of this century a whole, and not a splintered fragmentation of specialities.

Is There Life on Other Worlds?

Carl C. Kiess

Carl C. Kiess, *a native of Fort Wayne, Indiana, was born on October 18, 1887. In addition to local high school work, his training during summer vacations continued with student courses offered by manufacturers in the field of mechanical and electrical engineering. After graduating from high school with highest honors, he entered Indiana University in 1906. After four years, with astronomy and mathematics as major courses, physics and chemistry as minor courses, he was awarded the A.B. degree with high honors. He then matriculated at the University of California as a Fellow at the Lick Observatory where he engaged in all phases of the daily work of a large research observatory. The winter semesters were spent at the University at Berkeley attending lectures in theoretical astronomy, mathematics and physics. The Ph.D. degree was awarded in 1913. Following this, he held positions teaching astronomy at the University of Missouri, Pomona College, and at the University of Michigan. At the outbreak of the War in April 1917, he was invited to join the staff of the Optics Division of the National Bureau of Standards to engage in research on military problems. At the end of the war he was among those invited to remain permanently where he continued for the ensuing forty years as a member, first of the spectroscopy section of the Optics Division, and later, of the Division of Atomic Physics. During his later years at the Bureau of Standards, he was invited first by Fr. P. A. McNally, S.J., and later, by Fr. F. J. Heyden, S.J., to cooperate in the establishment of research work at the Georgetown College Observatory where he now teaches.*

Is There Life on Other Worlds?

FROM ANCIENT TIMES to the present the question "Is life there?" has greeted all explorers on their return, whether by land or by sea, or from the highest elevations or the lowest depths. The earliest explorers brought back tales of fabulous creatures—of dragons and griffins on land, and of mermaids and monstrous sea serpents in the oceans. At a later age, Columbus and his followers were more realistic and offered evidence in support of their discoveries in the form of new items of diet, such as the potato and Indian corn, and other gustatory delights, including tobacco. The age of geographical discovery is still with us, and whether it pertains to the highest elevations on land, or the lowest attainable depths of the sea the same question persists.

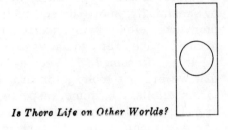

The age of space discovery began 350 years ago when Galileo devised his "optical tube"—as he called his telescope. With it he examined the Sun, the Moon, and the planets of our solar system, and the stars also. One evening when he was entertaining some of his friends, one of them asked after viewing Mars, "Do you believe life is there?" to which he replied, "I cannot say yes, but I cannot say no." And this question has been and is still being asked after each new advance and improvement in the art of astronomical observation. Today, we are in the midst of vast preparations to bypass earth-based telescopes and their accessories to get the answer to this age-old question by seemingly more direct methods of on-the-spot exploration, either with adequately instrumented or with manned space craft, and at vastly greater expense.

It is presumed, in seeking for extraterrestrial life, that we are not looking for dragons or mermaids, but for forms similar, in some respects, to those with which we are acquainted on our Earth. The biochemists tell us that life is a manifestation in various aspects of carbonaceous and amino compounds. These manifestations, or activities, are: genesis or birth, growth and development to maturity, reproduction in kind for continuation of the species, followed by decay and death for the individual. These activities are narrowly circumscribed by habitat and temperature. The activities of all living organisms must occur in a medium of water, supplemented for most of them by oxygen also. If deprived of water, the organism dies. If the temperature exceeds about 145°C the complex molecules, of which the organism is built, break asunder and the individual dies. If the temperature falls too low, again the individual may die because of freezing of its water, or it may relapse into a state of dormancy owing to lack of energy necessary for its activities. Some living organisms, including man, display a trait that we may designate as mind, wherewith an individual or a group of individuals may modify their environment so that locally they may survive the naturally imposed limitations; as for example with clothing, shelter, air conditioning, etc. Such artificial environments are used by our astronauts.

The activities of all living organisms on Earth depend directly or indirectly on radiant energy from the Sun—not only the sunlight that we see and the solar heat that we feel, but the invisible radiations in the infrared and ultraviolet regions beyond reach of our physiological modes of detection. All of these solar radiations constitute the Sun's spectrum of which the individual rays are specified by numbers called wavelengths. You may see the visible portion of this spectrum in the rainbow or in the spray of your garden hose. The invisible portion may be detected and measured either photographically or with electronic devices. For visual observation a better device than the rainbow is a glass prism. Although Newton was among the first to use it for studying the solar spectrum, it was the German optician Fraunhofer who first discovered features in this spectrum that, with others, have subsequently been fundamental in all astrophysical research.

In Fig. 1 we see the solar spectrum improved in scale over the primitive observation that Fraunhofer made more than 125 years ago. You will see it as he did, crossed by several dark gaps, or lines, from the deepest red to the violet end. At first he noted about ten of them which he labeled with the capital letters A to K. Later, with improved instruments he noted more, so that when his studies of the solar spectrum ended he listed about 125 lines, some of which he designated with small letters. Today, with our vastly improved and powerful ground-based spectrographs more than 30,000 of these lines have been photographed and catalogued. They appear in the spectra of the Moon, the planets, and other objects that are illuminated by sunlight. About a century ago the explanation of these lines was given by Kirchhoff and Bunsen, who interpreted them as the characteristic absorptions of the hot gases in the Sun's atmosphere. The Sun is so hot that nearly all the elements and some of the compounds of our earthly experience are in the vapor state and absorb some of the energy emitted by the Sun. If a planet or satellite in the solar system has an atmosphere then before the sunlight reaches its surface the atmospheric constituents of the planet will also absorb solar rays.

Fig. 1. Solar spectrum, Yerkes Observatory photograph.

Thus, the lines designated A, B, and a, by Fraunhofer are, as we now know, due to absorption by oxygen and water vapor in the Earth's atmosphere.

Figure 2 illustrates not only the dual nature of the spectrum of sunlight as we record it, but shows also another very fundamental principle of astrophysics, namely, the Doppler–Fizeau effect, whereby the position of the dark lines in a stellar spectrum depends on the velocity with which the emitting source is approaching or receding from the observer. The Sun rotates on its axis in about 25 of our days. Its eastern limb is approaching us, and its western limb is receding. Spectra from the two regions will show a jogged effect.

However, lines originating in the Earth's atmosphere will be unaffected. Thus, it is a unique way of separating

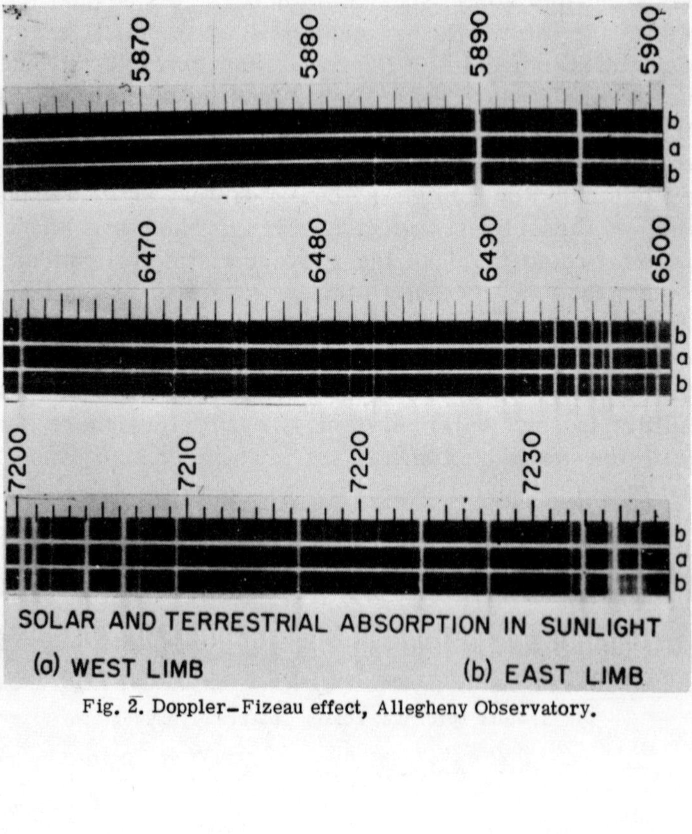

Fig. 2. Doppler–Fizeau effect, Allegheny Observatory.

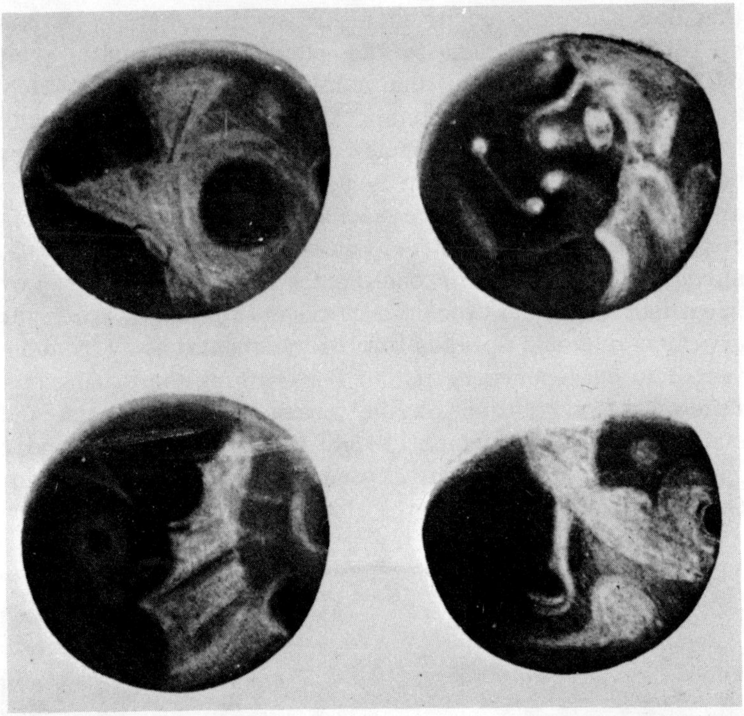

Fig. 3. Barnard's drawing of Mars, Lick Observatory photograph.

the absorption spectrum of a planet's atmosphere from that of the Sun.

Our search for life on other planets will be limited, therefore, to a search for water vapor and oxygen in their atmospheres. We may dismiss from consideration all the planets of our solar system except Mars because they show neither the temperature conditions nor the atmospheric properties necessary for living organisms of the kinds we know.

Figure 3 shows a drawing of Mars as it appeared about 70 years ago to one of our best observers at the telescope. You will note the polar caps, and the light and dark areas. From the shift of these surface features across the planet's disk a rotation period about 24.6 hours has been derived so that day and night occur much as on the Earth. The axis of rotation is inclined to the plane of the planet's orbit by about 24.5°—almost the same as the Earth's, which means that spring, summer, autumn, and winter occur much as on Earth, but they are about twice as long as ours because the Martian year is nearly twice ours; and they are relatively more severe because of the greater eccentricity of the Martian orbit. When the polar caps melt in the spring and summer of one hemisphere the dull gray or brownish color of the dark areas becomes green or greenish-blue and spreads toward the equator away from the receding cap—contrary to the behavior of the springtime renewal of vegetation on the Earth. At the sunrise and sunset limbs of the planet, bright clouds frequently appear and vanish in the regions exposed to full daytime sunlight. Occasionally another type of cloud is seen—diffuse, yellow

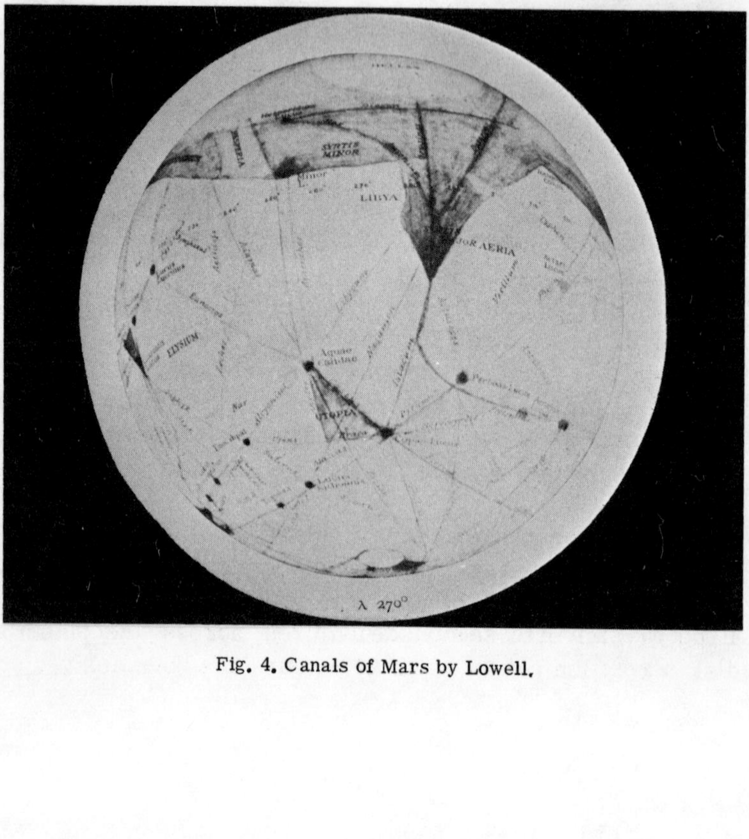

Fig. 4. Canals of Mars by Lowell.

clouds originating in the equatorial regions and spreading, sometimes with high speeds, into the temperate zones. They are more long-lived than the bright clouds and may persist for many days. They are commonly called dust storms.

Figure 4 shows one of the many drawings of Mars by Percival Lowell, who was one of our most prominent amateur astronomers. I use the word in its true meaning. In addition to the usual dark features the drawing shows many of the dark streaks that he observed and designated as canals—not as channels as designated by the Italian astronomer, Schiaparelli, who first saw them. Lowell interpreted them as artificial watercourses through farmlands irrigated and cultivated by a highly intelligent people who must conserve every drop of the precious fluid from the melting ice caps. The fact that not all the canals were equally visible in different years indicated to him that the Martians prorated the water among them so that enough would be available to meet the needs of the people in alternate years. You may read all about it in his charmingly written books that appeared in the early years of this century.

You may note that the interpretation of Mars thus given is based entirely on analogy with terrestrial features—polar caps of the snow and ice of water, green vegetation awakened by the warmth of moisture of spring and summer, cumulus and cirrus clouds in the morning and evening sky, and dust clouds blown from the deserts—all of them phenomena that occur because of water vapor in the atmosphere. Nevertheless during all the controversy about life on Mars no astronomer, for the past 70 years, has ever produced convincing proof from spectroscopic observations that water vapor is in the Martian atmosphere. Some have believed that they have done so; others, equally capable, have not. Yet, if the polar caps are frozen water, then, when one melts there must be a transport of water through the atmosphere because there are no seas or lakes on the Martian surface.

In a period of two years and two months Mars is

conspicuously visible in our midnight sky. It is then in opposition to the Sun. Not all oppositions are equally favorable for observing the planet; but those that occur when Mars is in that part of its orbit nearest the Sun are especially favorable because the planet is then within 35 million miles from the Earth. These favorable oppositions occur at intervals of 15 and 17 years. Such an opposition occurred in September 1956. Early in that year Dr. Lyman J. Briggs, Emeritus Director of the National Bureau of Standards, and at that time serving as Chairman of the Research Committee of the National Geographic Society, told me of the Society's project for sending a team of observers to a South African Observatory to study Mars during the weeks preceding and following its opposition. However, he said that he saw in the schedule of observations no reference to the water situation on Mars. "Has that been satisfactorily settled?" he asked. My reply was that it had not been nor would be settled until astronomers used a sufficiently powerful spectrograph to clearly separate the Martian water vapor spectrum from the Earth's. His next question was "How would you like to undertake it?". I asked for a few weeks delay before giving an answer. So I went into consultation with my colleague, Mr. Charles Corliss, at the National Bureau of Standards, and together we went to Father F.J. Heyden, S.J., of the Georgetown College Observatory and obtained from him permission and help in using the facilities there for some preliminary tests. The upshot of it was that the end of June 1956 found us at the recently completed Slope Observatory near the summit of Mauna Loa, Hawaii. This facility was established by the U.S. Weather Bureau which accorded us the privilege of being the first to use it. We had with us parts of a powerful spectrograph supplied by the Bureau of Standards, and were supported financially by the National Geographic Society, to try out on Mars the experience that we had gained at the Georgetown observatory.

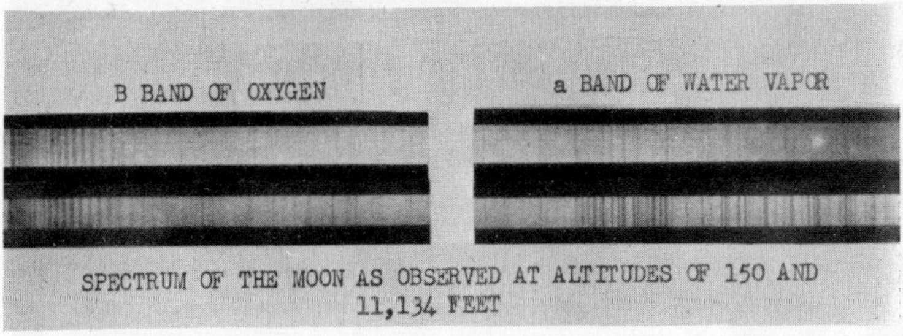

Fig. 5. Effect of H_2O spectrum of the Moon at Georgetown and at Mauna Loa, Georgetown University Observatory and Mauna Loa Observatory.

Figure 5 contains sections of the Moon's spectrum photographed at Georgetown Observatory and at Mauna Loa, where we were more than two miles above sea level, and a mile above the bulk of the water vapor in the atmosphere. Although the strength of the oxygen lines is about the same at the two stations, the effect of altitude on the absorption of the water vapor is clearly evident.

Figure 6 shows a section of one of our best spectrograms. In color these spectra would appear red; light red at the top left, deeper red at the lower right. From this strong line, marked Hα, which is due to hydrogen in the Sun, you can see that the Martian spectrum is shifted relative to the Moon's. The middle strip is in the region of Fraunhofer's "B" band of oxygen. Careful inspection of it reveals no lines in the Martian spectrum accompanying the terrestrial lines. The bottom strip shows the region of Fraunhofer's "a" band, due to water vapor. Again no Martian lines are present. The mark near 6985 Å indicates a pair of terrestrial lines that appear with different relative strengths in the lunar and Martian spectra. This results from a shift of a masked line of iron in the Moon's spectrum to coincidence with the atmospheric line in the Martian spectrum.

Because of the negative results yielded by our observations on Mauna Loa we packed up our equipment and came home. In telling our story to some of our fellow

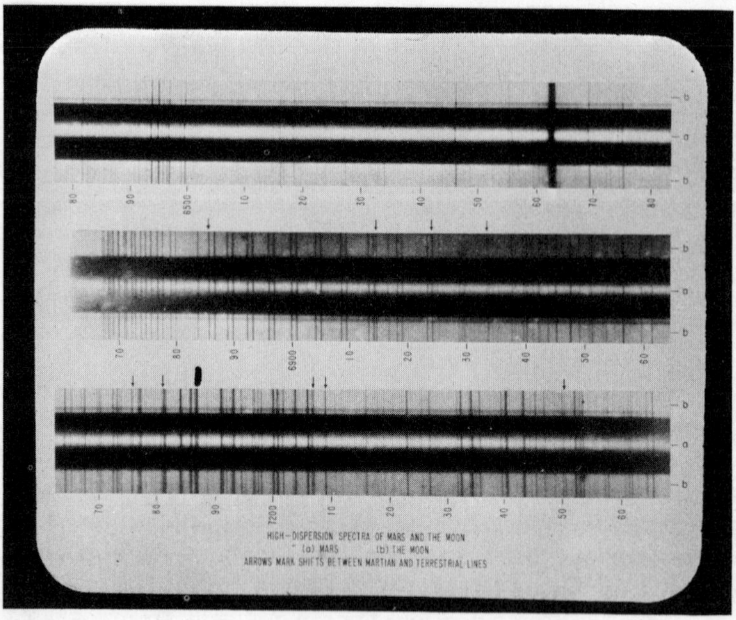

Fig. 6. Mark near 6985 A shows pair of terrestial lines that appear with different relative strengths in Lunar and Martian spectra, Georgetown University Observatory, with permission of the Astrophys. J.

spectroscopists one of them suggested that in the light of our experience we set up our instruments again to observe the near infrared region of the Martian spectrum later in the season when Mars would be moving away from the Earth. We would then be able to record Fraunhofer's "A" band of oxygen and the water vapor band beyond the "a" band. We did this, again at Georgetown Observatory and with the support of the National Geographic Society. Again our results were negative as far as oxygen and water vapor were concerned.

Now it happened that in the interim between our work on Mauna Loa and at Georgetown University a remarkable event took place on Mars, of which we were unaware at the time. After the third week in August a yellow cloud began to form on the planet. In a few days it spread over the entire surface with a deep yellow pall that hid the surface features, including the polar caps, from view. No event quite like this had ever been reported before in the known history of Mars. Observers of it in various parts of the world state that the color deepened to dark yellow, to yellowish-brown, to reddish-brown, during the later stages of the cloud, which all the observers except a few have designated as a "dust" storm of unusual extent and duration. When, after a few weeks, the clouds broke, the surface features, especially the polar caps, appeared much as they usually do, unsullied by layers of dust. One eminent astronomer has suggested that the vegetation on Mars has very rapid recuperative powers and shot up green stems and leaves above the dust cover. Another eminent astronomer, not impressed with this explanation, has suggested that gentle breezes blew the dust into crevices and cracks on the surface.

We have a different explanation. One of our best observations was made on the night of October 11-12, 1956, when observers elsewhere noted that Mars was still veiled with thin remnants of the great storm. One pair of these observers described the veil as distinctly reddish-brown in color. At the time of our observation we were unaware of the fact that the light recorded by our spectrograph had passed twice through this obscuring substance.

When a geodesist wishes to portray different elevations over an area of the Earth's surface he draws a contour map. When he wishes to portray differences in elevation relative to a datum in a given direction, say an E-W line he draws a profile map. The astrophysicist does a similar thing with an instrument called a microphotometer. His datum level is the continuous background of sunlight, or starlight, on which appear as dips and depressions the absorption features of the atoms and molecules in the gaseous atmospheres through which the light passes on its way to the observer.

Figures 7, 8, and 9 are reproductions of microphotometer tracings of our spectra of Mars and the Moon obtained on this favorable night of October 11-12, 1956. The reader will see the differences between them. The more conspicuous ones are marked with arrows that indicate prominent features in the absorption spectrum of nitrogen dioxide, NO_2.

To assure ourselves that this was not a casual or temporary phenomenon, we again observed the Moon and

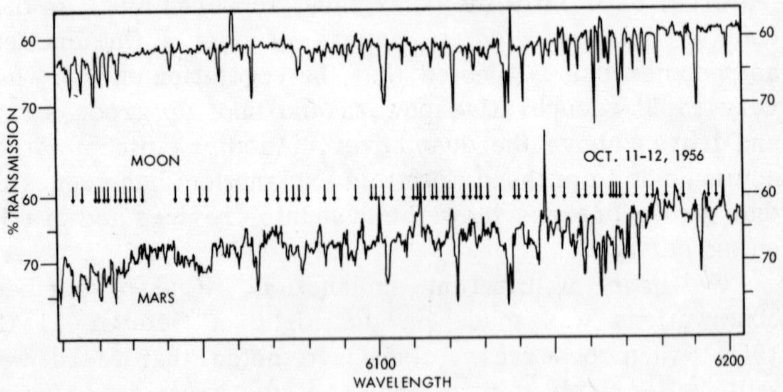

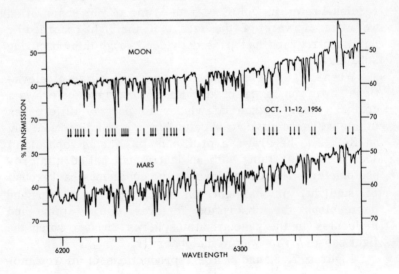

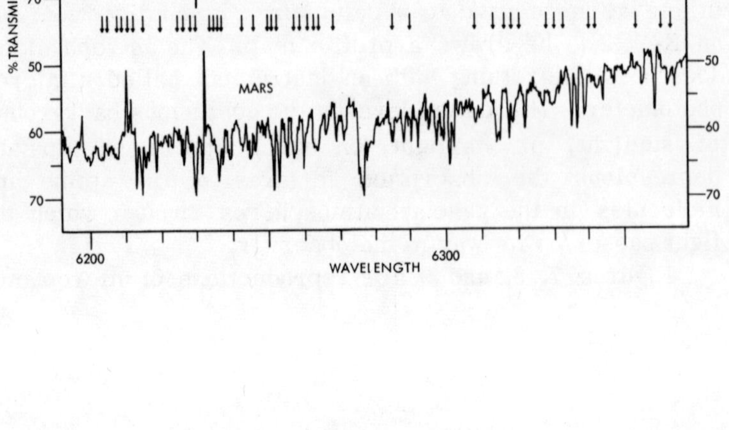

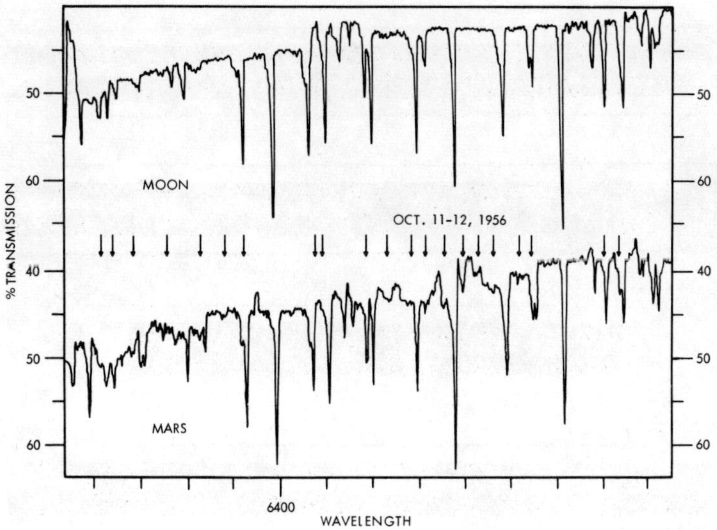

Fig. 7, 8, and 9. Microphotometer tracings of Lunar and Martian Spectra, Georgetown University Observatory.

Mars when the planet was in opposition late in 1960. Figure 10 shows sections of the spectra of Mars and the Moon from the D lines of sodium to the H and K lines of calcium in the violet. It portrays the striking falling off in intensity of the Martian spectrum with decreasing wavelength. Particularly noticeable is the depression near the "b" band of Fraunhofer, which we now know is due to absorption by magnesium. The differences between the spectra of the Moon and Mars, are noticeable, and again they agree with absorption features of NO_2 revealed by laboratory experiments and illustrated by the microphotometer tracings in Figure 11.

About a year ago Mars was again in opposition. The same prominent absorption features of NO_2 in the blue region of the spectrum of Mars appeared again at this time.

Now what kind of a gas is NO_2? The chemistry books tell us that it is a yellow gas, the degree of yellowness depending on its concentration, which in turn, depends on

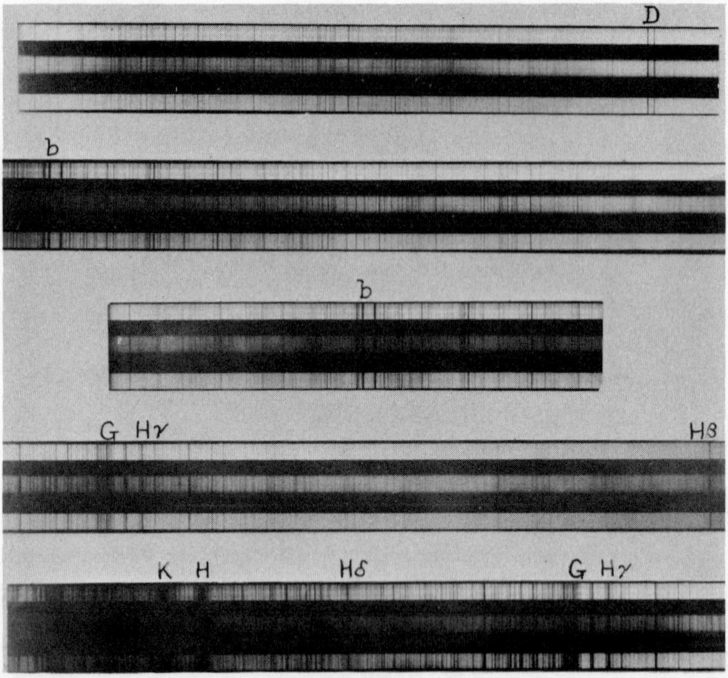

Fig. 10. Section of the spectra of the Moon and Mars, Georgetown University Observatory. In each strip the spectrum of Mars is between the two Lunar spectra.

its temperature. At ordinary room temperatures, it constitutes about 20% of a mixture of which the major constituent is its dimer nitrogen tetroxide, N_2O_4, a heavy transparent gas that freezes to a white, crystalline snow or ice at -11°C. Figure 12 shows the yellow NO_2 molecules in the shape that physical chemists have found for them. The right side shows also how they join together when the temperature falls. At this stage the gas is transparent. When sunlight shines on the mixture some of them break apart to form NO and O, both transparent to visible light. At low temperatures the NO may link up with NO_2 to form N_2O_3 which is a blue liquid, or it may be frozen in with the N_2O_4 and impart to it a bluish tint; or if the mixture melts and wets the minerals of the soil, it may color them green. All of these color characteristics of this very versatile family of oxides of nitrogen that unite and dissociate in response to local changes in temperature and pressure in the Martian atmosphere fit the color patterns that two or more generations of astronomers have described for the clouds, the polar caps, and the dark areas of Mars.

The medical books on toxicology tell us that N_2O_4 and NO_2 are poisonous. In our atmosphere small amounts of about 25 molecules per hundred million of nitrogen and oxygen molecules cause irritation to our eyes and respiratory passages. In considerably larger amounts they are fatal. The smaller amounts that occur in the smog of some of our westernmost areas are already causing

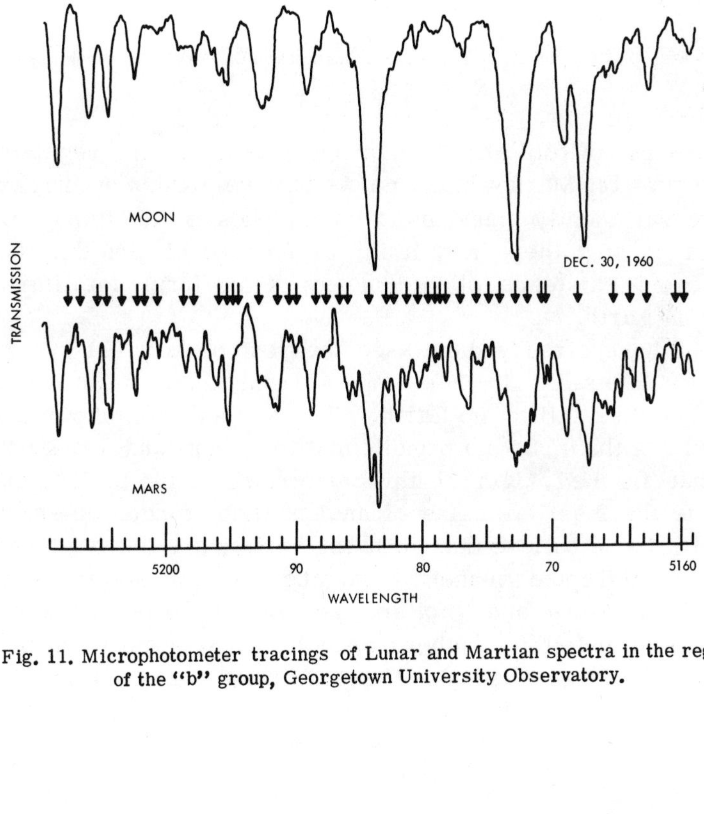

Fig. 11. Microphotometer tracings of Lunar and Martian spectra in the region of the "b" group, Georgetown University Observatory.

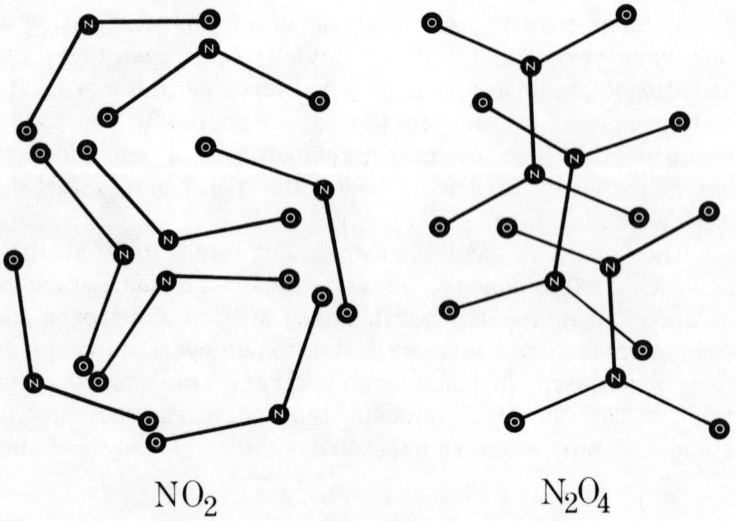

Fig. 12. NO_2 and N_2O_4 molecules, Georgetown University Observatory.

millions of dollars of damage to fruit and vegetable crops. On Mars where, as we believe, oxides of nitrogen are sufficiently abundant to form clouds in the atmosphere and caps at the poles, it is our firm conclusion that they prevent existence of life in any of the forms familiar to us on Earth.

I have already dismissed the rest of the solar system as unfavorable to life. Yet from time to time stony meteorites fall to the Earth. Chemical analysis shows that a few of them contain organic matter—compounds of carbon. What is the origin of this material? Some believe that it is the fossil remains of ancient living forms, others do not. Some believe that these meteorites are the fragments of a disrupted planet, a member of our solar system between Mars and Jupiter, once the abode of life before disaster befell it. Others do not. Some believe that they are the fragments of a life-sustaining planet, later disrupted, of another solar system. But this too is hypothetical. So the question is an open one, and may serve as an excuse, or an objective, for other adventures into the depths of space.

Outside our solar system, but within the galactic system to which we belong, there are hundreds of thousands, if not millions, of stars of exactly the same physical and chemical state as our Sun. Would it not be strange indeed if our Sun were the only one of its fellows with a retinue of planets? There is evidence that some of them are accompanied by dark companions. And would it not be stranger still if among the many planetary families that probably exist, not only in our own galactic system but in the tens of thousands of other galaxies, our Earth were the only abode of life? Most astronomers, who have expressed opinions on the subject think it highly probable that there are many more Earths in the Grand Universe, all endowed with life. Perhaps on one or more of them, at this very moment, there are beings wondering what we are doing here.

Plans to send or receive communications from these external worlds are already in consideration, if not actually begun, by our radio astronomers. However the only way we know of now for exchanging messages is by means of electromagnetic waves—in other words with light or radio waves that travel with a speed of 186,000 miles per second. The round trip for a message from us to the nearest star would require about ten years, if the intelligent beings there could decode our message and send back a reply. To other stars, Capella for example, it would require about 90 years. How many of us would live long enough to know how happy our distant relatives were to receive announcements of our birth?

The question with which we began this discourse is still open, and to it we can give only the Galilean answer: "We cannot say Yes, but we cannot say No."